T0191962

Grape Rootstocks and Related Species

Alireza Rahemi · Jean C. Dodson Peterson ·
Karl True Lund

Grape Rootstocks
and Related Species

 Springer

Alireza Rahemi ⓘ
College of Agriculture
Virginia State University
Petersburg, VA, USA

Karl True Lund ⓘ
Cooperative Extension
University of California
Madera, CA, USA

Jean C. Dodson Peterson ⓘ
Department of Wine and Viticulture
Department of Horticulture and Crop
Science
California Polytechnic State University
San Luis Obispo, CA, USA

ISBN 978-3-030-99409-9   ISBN 978-3-030-99407-5   (eBook)
https://doi.org/10.1007/978-3-030-99407-5

This Springer imprint is published by the registered company Springer Nature Switzerland AG
The registered company address is: Gewerbestrasse 11, 6330 Cham, Switzerland

Foreword

Grapevine rootstocks are an essential feature for sustainable viticulture around the world. Only a few grape-growing regions are spared infestation by phylloxera, and pests and diseases of grapevine roots challenge production nearly without exception. Since the introduction of phylloxera into Europe, the subsequent eruption of phylloxera in European viticulture, and the distribution of phylloxera very widely, the value of rootstocks in viticulture has been firmly appreciated and rootstocks steadily and increasingly recognized as critical for successful viticulture. Those who seek to maximize vine performance through rootstocks should understand the attributes and origin of rootstocks and how the characteristics of rootstocks may influence scion performance and vine adaptation.

This book describes the importance, utilization, and future of rootstocks in the context of grapevine botany, plant breeding, and horticulture. The early years of the phylloxera crisis saw pioneering research with many North American grape species selections and their hybrids, both as rootstocks and in direct production. That research was fueled by crisis and the exigency of identifying practical approaches for phylloxera management. Rootstocks developed in that first wave of research are still in use, and the fundamental biological and viticulture knowledge of phylloxera/rootstock/scion interactions guides further developments and investigations. Yet now, a more comprehensive understanding of *Vitis* species, their origins, relationships, and attributes informs viticulture, grape breeding, and the utilization of rootstocks.

We recognize that the North American grape species that coexist with phylloxera in nature show adaptations to reduce phylloxera damage, such as resistance and tolerance to the pest. Many questions about the ecology and evolution of grapevines and their pests and pathogens are relevant to species research and rootstock utilization and breeding. How does natural gene flow among grape species and populations influence interactions with phylloxera and other pests? Why does resistance and tolerance against pests and diseases with a delimited original distribution exist in grape species native to other areas? How stable is the resistance and tolerance of our rootstocks against pests and diseases that must infest to order to survive? How does the use of rootstocks shift pest and disease populations? In vineyards, phylloxera and other pests often encounter a genotypical monoculture, yet in nature, dioecy and vegetative propagation in this long-lived liana engender a mosaic of genetic diversity.

The chief role of rootstocks in viticulture is to provide protection against pests and diseases of the vine roots, with phylloxera protection the keystone. Other pests and pathogens, including fungi, bacteria, and nematodes, have long been recognized as important threats, and rootstocks that protect against them are increasingly the goal of breeding and evaluation programs. Is phylloxera vanquished? Hardly! Rather, protection against other pests and pathogens is taking its place alongside phylloxera protection as pest and pathogen populations shift with changing environments, management practices, and production regions. We now have the germplasm and evaluation methods to develop and introduce rootstocks that provide protection against crown gall and ground pearls, and more pest and disease protection is revealed as investigations continue. We can and must reduce pesticide applications to vineyards by using protective rootstocks instead.

Rootstocks have an important and emerging place in the management of grapevine virus diseases. Fanleaf degeneration and other diseases associated with nematode-transmitted viruses now may be managed with rootstocks that are resistant to nematodes or that provide protection against the diseases themselves. Yet, the mechanism and genetic drivers of this protection against virus diseases are poorly understood, which slows progress in breeding and introducing rootstocks with adaptation to diverse soils and climates. Rootstocks influence the development of leafroll disease and other virus-associated grapevine diseases, and the deployment of leafroll disease resistance and tolerance in rootstocks that provide protection against phylloxera and other pests would be an important step in harm reduction. Rootstocks and their progenitor species will be a source of genetic resistance and tolerance against virus disease protection for scion variety improvement.

While a handful of rootstock varieties dominate the global grapevine nursery trade, extensive and diverse breeding and evaluation programs continue to create and characterize rootstocks with superior adaptation to specific environments and varieties and which provide protection against newly emerging pests and diseases as well as against recognized threats. The movement of grape plant material in a secure phytosanitary context is an important contributor to the adoption of rootstocks. Similar growing conditions may favor the utilization of the same rootstocks in different regions for similar benefits, and familiarity both with well-established rootstocks and new rootstock variety introductions is essential. Shared trends drive shared needs–water use efficiency, yield efficiency, sodium and chloride tolerance, and adaptation to acidic and calcareous soils are valued in many areas, and rootstocks that provide special value must be introduced widely.

Through our familiarity with the species, their phenotypic characteristics, and their horticultural contributions increase steadily, new frontiers and opportunities are presented by technologies such as recombinant DNA technology, gene editing, somatic hybridization, doubled haploids, and synthetic seeds that are as yet incompletely realized in rootstock improvement and utilization. Botanical exploration and analysis reveal distinctions and relationships among species and populations, with newly recognized distinct types to be expected even in well-described habitats and with new Neotropical species on the horizon. From this book, future initiatives in grape rootstock breeding, horticulture of grape rootstock evaluation and use, and grape species research will be inspired.

Modesto, California, USA

Dr. Peter Cousins
Grape Breeder and Geneticist

Preface

The genus *Vitis* includes approximately 80 different species of grapevines, depending on the subtle shifts in taxonomic classification taking place as natural hybridization and ongoing classification efforts continue to evolve. Of all species in the genus *Vitis*, the European grape, *Vitis vinifera*, is certainly the most important worldwide for the commercialized cultivation of table, raisin, and winegrapes. Grape species other than *V. vinifera* have also historically been used in the making of juices, wine, jellies, and jams. Many of these wild grape species have nutraceutical properties and health benefits that have drawn consumer attention and contributed to interest in incorporating them into breeding programs beyond the traditional efforts of rootstock hybridization from wild species.

Initially, *V. vinifera* was propagated asexually and without the common day practice of grafting to wild species hybrids. In the late nineteenth century, an aphid-like pest, phylloxera (*Daktulosphaira vitifoliae*), an aphid-like pest, was inadvertently introduced from North America to Europe and ultimately impacted most grape-growing regions worldwide. Phylloxera causes damage by feeding on the structural roots of *V. vinifera*, eventually leading to vine death. In an effort to save the wine industry, American *Vitis* species which coevolved with phylloxera were eventually used as rootstocks for the widely planted *V. vinifera* scion cultivars. The use of wild species directly as a rootstock has become a common practice. Later, crosses between these wild species became common, and historical breeding efforts led to the modern grape rootstock breeding programs currently active around the world. Today, grape rootstocks play a fundamental role in resistance to biotic and abiotic stresses and adaptation of grapevine to different environmental conditions, a factor that has opened commercial grape growing up to regions that might otherwise be overlooked. Grape rootstocks can be used for adaptation to a variety of soil conditions, including soil texture, depth, nutrient availability, pH, salinity, lime content, water availability (drought), and water drainage. Rootstocks can also be used to shift scion cultivar the timing of various key phenological events and indirectly affect vineyard design.

Regardless of the specific rootstock selected or the environmental challenge addressed, it is necessary to introduce new rootstocks using resistance advantages from different species. There are around 1500 grape rootstocks developed in the world, of which around 50 are commonly used as commercial rootstock. North American species account for around 30 species, and two-third of them have already been used for rootstock breeding. However, it has been reported that the most commonly available rootstocks are derived from just three American species *(V. berlandieri, V. rupestris,* and *V. riparia).* Therefore, the most common grape rootstocks have a narrow genetic base, and efforts to extend the gene pools for breeding programs by using the other species are of ongoing importance to the industry and scientific community.

This book will annotate about 20 grape species that are vitally important in breeding programs and provide information on approximately 150 of the most familiar grape rootstocks in the world.

A great deal of research has been performed on grape rootstocks around the world. However, the resources and information are scattered, inaccessible, and unexploitable for the vast majority of grape growers and researchers. In this book, we have made substantial efforts to gather and classify this information and combine it with our own opinions and experiences in the hope that it will be useful to those interested and accessible to a broad audience. Various sources included in the text required permission and copyright approval. As such, we spent substantial time on this critical issue. It is necessary to thank all the authors, researchers, institutions, and associations for their help and cooperation. Some of the resources were particularly difficult to source and it would not have been possible to include them without the support of the corresponding authors and their contacts. This book is the result of the dedicated efforts of all of them. We hope that all faculties, researchers, and specialists will help us improve the quality of the book in future editions by sending comments to us.

Petersburg, Virginia, USA   Alireza Rahemi
San Luis Obispo, California, USA   Jean C. Dodson Peterson
Madera, California, USA   Karl True Lund

Contents

About the Authors

Dr. Alireza Rahemi is Postdoctoral Research Associate at Virginia State University, USA. He has done extensive research and continues to work on wild relatives of subtropical and temperate fruits and nut trees. He worked as a Horticulture Specialist at the Ministry of Agriculture in Iran. He received his Ph.D. in Horticulture and conducted part of his research at University of Florida, USA. His Ph.D. research was focused on wild almond species, including genetic diversity using new emerging molecular techniques. He has also conducted research on the wild grape species of Southern Ontario and their utilization as rootstock during his postdoctoral research at the University of Guelph, Canada.

Dr. Jean C. Dodson Peterson is Associate Professor of Viticulture at California Polytechnic State University, San Luis Obispo. She earned her undergraduate degree from the University of California, Davis, in Viticulture and Enology with a minor in Managerial Economics. She continued at UC Davis, earning her M.S. and Ph.D. in Horticulture and Agronomy. She teaches basic viticulture, the advanced viticulture series, advanced independent study, senior project research, and thesis courses. She has been awarded the Panhellenic Award for Excellence in Education and the Terrance Harris Award for Excellence in Mentorship. She has also earned her Certification in Effective College Instruction from the Association of College and University Educators and the American Council on Education. Her research program strives to advance the understanding of grapevine physiology, rootstock–scion interactions, rootstock selection, and vine performance factors. Her research mission is to execute innovative and inspiring research that advances the wine and viticulture industry and scientific community.

Dr. Karl True Lund is University of California Viticulture Advisor for Madera, Merced, and Mariposa Counties. The primary research he is currently conducting is looking at the adaptation of nematode-resistant rootstocks for high yielding mechanized vineyards. Prior to his current position, he was Lecturer at California Polytechnic State University San Luis Obispo, where he taught introductory viticulture and viticulture pest management classes. He conducted his Ph.D. work at the University of California at Davis. His Ph.D. work looked at the diversity of phylloxera within California. He also assessed the resistance of wild accessions and current rootstocks to the different types of phylloxera he identified within the state.

Chapter 1
Introduction

Grape cultivation is considered to be as old as civilization (OIV 2017). Today, grape-derived products are one of the most economically important fruit crops (Wen et al. 2018), with more than 10,000 grape varieties grown in 100 countries, on an estimated 7.5 m ha, with approximately 75 m tons of production each year (OIV 2017). Most production occurs in the zone between 30° and 50° latitude in the northern and southern hemispheres (Stevenson 2005).

Around 50% of world grape production is used to make wine (wine grapes), with 33% eaten as fresh fruit (table grapes), and the remaining 17% used as raisins, juice, or stored in the form of grape musts (freshly crushed fruit) whether concentrated or not (FAO-OIV 2016; OIV 2017). Most commercial production is with pure *Vitis vinifera* (Christensen 2003). 'Kyoho', 'Cabernet Sauvignon', 'Sultanina', 'Merlot', 'Tempranillo', 'Airen', 'Chardonnay', 'Syrah', 'Red Globe', 'Garnacha Tinta'/'Garanche noir', 'Sauvignon blanc', 'Pinot noir'/'Blauer burgunder', and 'Trebbiano toscano'/'Ugni blanc' are, in sequence, the varieties with the highest production and they cover more than one-third of the planted area (OIV 2017). 'Kyoho', a Concord-like cross popular in East Asia, is the most commonly planted variety by acreage (OIV 2017). It is not pure *V. vinifera*; however, it still owes more than 50% of its parentage to *V. vinifera* (VIVC 2019). The initial reason for the wide adoption of rootstocks was, and in many locations still is, due to phylloxera (*Daktulosphaira vitifoliae*), an aphid-like pest of grapevines. Phylloxera is native to the Eastern portion of North America and was inadvertently imported into the wine-producing regions of France in 1860 (Skinkis et al. 2009). By the end of the nineteenth century, France had lost more than 70% of its wine production (Robinson and Harding 2016). The French wine industry was historically based on the exclusive use of own-rooted *V. vinifera*, which has no resistance to root-feeding by phylloxera. However, some American *Vitis* species had co-evolved with phylloxera and thus had evolved tolerance to the pest. The use of hybrids between French *V. vinifera* varieties and American *Vitis* species failed to produce a variety that was resistant to phylloxera while also being considered to produce wine of commercially acceptable quality. The French wine industry eventually succeeded in addressing phylloxera by grafting

© The Author(s), under exclusive license to Springer Nature Switzerland AG 2022
A. Rahemi et al., *Grape Rootstocks and Related Species*,
https://doi.org/10.1007/978-3-030-99407-5_1

French wine varieties (as scions) onto an American *Vitis* species, or species hybrid, as a rootstock. As phylloxera spread from France to virtually every wine-growing region worldwide, so did the use of rootstocks (Campbell 2004). The first rootstocks were simply selections of individual wild American *Vitis* species, some of which are still used today. The two initial American species utilized as rootstocks were *Vitis riparia* and *Vitis rupestris*. Rootstocks derived from these two species were initially successful but eventually declined in France's lime-rich soils as neither tolerated calcareous conditions. As a result, *Vitis berlandieri* was collected from America and incorporated into rootstock breeding programs in an effort to convey lime tolerance. Lime, however, was not the only abiotic soil condition that rootstocks must address. Over time, soil characteristics such as texture, drainage, salinity, and pH have all been addressed or attempted to be addressed through rootstock selection. Furthermore, biotic issues other than phylloxera have also required the use of rootstocks to address properly. Specifically, the control of nematodes has fallen mainly to rootstocks for control, an issue that is complex given the vast array of resistance levels across the various commercial rootstocks (Ferris et al. 2012).

In addition to dealing with issues in the soil and soil pests, rootstocks have been shown to affect scion growth (vigor) and yield (Keller et al. 2008; Kidman et al. 2014b; Nelson 2015; Dodson Peterson and Walker 2017), nutrition status (Holzapfel and Treeby 2007; Lambert et al. 2008; Kidman et al. 2014a), pollination, and fertilization efficacy (Kidman et al. 2014a), and berry characteristics and fruit chemistry (Rühl et al. 1988; Koundouras et al. 2009; Migicovsky et al. 2021). Expansion of grape production into new areas, development of new cultivars, and increasing concerns about adapting grapevines for changing climates continue to necessitate the use of wild grapevine species in breeding programs (Klein et al. 2018). An understanding of the effects a rootstock will have on a scion must start with an understanding of species parentage. Each American *Vitis* species, as well as the other grape species used in rootstock breeding, has a specific set of traits. They are possessing a basic understanding of the traits particular species have or are likely to have assisted in informing what traits might be observed in the newly bred progeny. The majority of rootstocks in use today are hybrids of three species: *V. berlandieri, V. riparia,* and *V. rupestris* (Galet 1998; Whiting 2005; Ollat 2016). Riaz et al. (2019) used 21 nuclear and 14 chloroplast markers to show that the relationship of many of these rootstocks is even closer. Twenty-six of the common rootstocks bred between the 1890s and 1930s had at least one parent coming from only three specific selections: *Vitis berlandieri* cv. Rességuier 2, *V. rupestris* cv. du Lot and *V. riparia* cv. Gloire de Montpellier. Overall, the results indicate that 39% of the genetic background of the major analyzed rootstocks originated from *Vitis berlandieri* cv. Rességuier 2, *V. rupestris* cv. du Lot and *V. riparia* cv. Gloire de Montpellier. The results found that the world's existing rootstocks have a narrow genetic base derived from only a few American grape species (Riaz et al. 2019). Lesser-known components of grapevine evolutionary biology include relationships among species that are used as rootstocks (e.g., *V. cinerea* var. helleri, *V. riparia, V. rupestris*). Hybrids between *V. cinerea* var. helleri and *V. riparia* or *V. rupestris* have been used to produce rootstocks that are easy to propagate, and that can withstand challenging abiotic conditions (Galet

1979). Single nucleotide polymorphisms (SNPs) used for a phylogenomic study in *Vitis* demonstrate that these important rootstock species occur in different clades within the North American subgenus *Vitis*. *V. cinerea* is most closely related to *V. vulpina*, while *V. riparia* and *V. rupestris* form a clade together with *V. acerifolia* (Miller et al. 2013). In general, the common grape rootstocks used today encompass around 20 different grape species. It is here where the industry must focus and advance understanding in order to make the most progress with future breeding efforts.

References

Campbell C (2004) Phylloxera: how wine was saved for the world. Harper Collins

Christensen LP (2003) Rootstock selection. In: Bettiga LJ (ed) Wine grape varieties in California. University of California, Oakland, CA, USA, pp 12–15

Dodson Peterson JC, Walker MA (2017) Influence of grapevine rootstock on scion development and initiation of senescence. Catalyst: Discovery into Practice 1(2):48–54. https://doi.org/10.5344/catalyst.2017.16006

FAO-OIV, Food and Agriculture Organization of the United Nations—International Organisation of Vine and Wine (2016) Table and dried grapes. FAO-OIV Publication. p 64. ISBN: 978-92-5-109708-3. http://www.fao.org/3/a-i7042e.pdf

Ferris H, Zheng L, Walker MA (2012) Resistance of grape rootstocks to plant-parasitic nematodes. J Nematol 44(4):377–386

Galet P (1979) A practical ampelography, Translated and adapted by Lucie T. Morton. Ithaca, NY: Comstock Cornell University Press

Galet P (1998) Grape varieties and rootstock varieties. Oenoplurimedia press

Holzapfel BP, Treeby MT (2007) Effects of timing and rate of N supply on leaf nitrogen status, grape yield and juice composition from Shiraz grapevines grafted to one of three different rootstocks. Aust J Grape Wine Res 13:14–22

Keller M, Smithyman RP, Mills LJ (2008) Interactive effects of deficit irrigation and crop load on Cabernet Sauvignon in an arid climate. Am J Enol Vitic 59:221–234

Kidman CM, Dry PR, McCarthy MG, Collins C (2014a) Effect of rootstock on nutrition, pollination and fertilisation in 'Shiraz' (*Vitis vinifera* L.), Vitis 53(3):139–145

Kidman CM, Mantilla SO, Dry PR, McCarthy MG, Collins C (2014b) Effect of water stress on the reproductive performance of Shiraz (*Vitis vinifera* L.) grafted to rootstocks. Am J Enol Vitic 65(1):96–108

Klein LL, Miller AJ, Ciotir C, Hyma K, Uribe-Convers S, Londo J (2018) High-throughput sequencing data clarify evolutionary relationships among North American *Vitis* species and improve identification in USDA *Vitis* germplasm collections. Am J Bot 105(2):215–226

Koundouras S, Hatzidimitriou E, Karamolegkou M, Dimopoulou E, Kallithraka S, Tsialtas JT, Zioziou E, Nikolaou N, Kotseridis Y (2009) Irrigation and rootstock effects on the phenolic concentration and aroma potential of *Vitis vinifera* L. cv. 'Cabernet Sauvignon' grapes. J Agric Food Chem 57:7805–7813

Lambert JJ, Anderson MA, Wolpert JA (2008) Vineyard nutrient needs vary with rootstocks and soils. Calif Agric 62(4):202–207

Migicovsky Z, Cousins P, Jordan LM, Myles S, Striegler RK, Verdegaal P, Chitwood DH (2021) Grapevine rootstocks affect growth-related scion phenotypes. BioRxiv 864850:29p. https://doi.org/10.1101/864850

Miller AJ, Matasci N, Schwaninger H, Aradhya MK, Prins B, Zhong GY, Simon C, Buckler ES, Myles S (2013) *Vitis* phylogenomics: hybridization intensities from a SNP array outperform genotype calls. PLoS ONE 8(11):e78680. https://doi.org/10.1371/journal.pone.0078680.g001

Nelson CC (2015) Zinfandel productivity is managed by pruning systems, applied water amounts, and rootstocks in a hot climate. PhD dissertation

OIV, International Organisation of Vine and Wine (2017) Distribution of the world's grapevine varieties. OIV Publication, p 54. ISBN: 979-10-91799-89-8. http://www.oiv.int/public/medias/5888/en-distribution-of-the-worlds-grapevine-varieties.pdf

Ollat N, Peccoux A, Papura D, Esmenjaud D, Marguerit E, Tandonnet JP, Bordenave L, Cookson SJ, Barrieu F, Rossdeutsch L, Lecourt J, Lauvergeat V, Vivin P, Bert PF, Delrot S (2016) Rootstocks as a component of adaptation to environment. In: Gerós H, Chaves MM, Gil HM, Delrot S (eds) Grapevine in a changing environment: a molecular and ecophysiological perspective. https://doi.org/10.1002/9781118735985.ch4

Riaz S, Pap D, Uretsky J, Laucou V, Boursiquot JM, Kocsis L, Walker MA (2019) Genetic diversity and parentage analysis of grape rootstocks. Theor Appl Genet, Under Rev, TAAG-D-18-00569

Robinson J, Harding J (2016) The oxford companion to wine, 4th edn. Oxford University Press

Rühl EH, Clingeleffer PR, Nicholas PR, Cirami RM, McCarthy MG, Whiting JR (1988) Effect of rootstocks on berry weight and pH, mineral content and organic acid concentrations of grape juice of some wine varieties. Aust J Exp Agric 28:119–125

Skinkis P, Walton V, Kaiser C (2009) Grape phylloxera, biology and management in the pacific Northwest, Oregon state university, extension service, EC 1463-E. Revised, p 25

Stevenson T (2005) The Sotheby's wine encyclopedia. Dorling Kindersley, pp 14–15. ISBN 0-7566-1324-8

VIVC, Vitis International Variety Catalogue (2019) Passbord data. Utilization: Rootstock. http://www.vivc.de

Wen J, Harris AJ, Kalburgi Y, Zhang N, Xu Y, Zheng W, Ickert-Bond SM, Johnson G, Zimmer EA (2018) Chloroplast phylogenomics of the new world grape species (*Vitis*, Vitaceae). J Syst Evol 56(4):297–308

Whiting JR (2005) Grapevine rootstocks, in Viticulture. Volume 1: Resources. In: Dry PR, Coombe BG (eds) Winetitles Pty Ltd, Ashford, Australia, pp 167–188

Chapter 2
Grape Species

2.1 Grape Species' Taxonomy

Vitaceae is a family with 14 genera and circa 900 species, including grapes (Christenhusz and Byng 2016). Grapes belong to the *Vitis* genus, which includes about 80 species (The plant list 2013) and is considered as two sub-genera, including *Euvitis* and *Muscadinia* (Muscadine grapes are sometimes considered as a separate genus in different classifications systems. The genus *Muscadinia* includes three species: *M. rotundifolia*, *M. munsoniana*, and *M. popenoei*). The *Euvitis* sub-genus includes the most important grape varieties. *Euvitis* is divided into three groups (Fig. 2.1).

1. The American group consisting of about 30 species that are important in rootstock breeding;
2. The Asian group, which consists of about 50 species, which up to this point have been less important for grape cultivation; and
3. The Eurasian group consists of a single species, *Vitis vinifera* L., which forms most of the different types of cultivated varieties in the world. There are two sub-species of *Vitis vinifera*: *Vitis vinifera* subsp. *sylvestris*, which is considered to be the wild form of grape; and *Vitis vinifera* subsp. *vinifera*, which refers to cultivated forms (OIV 2017).

To understand the relationships among grape species, the single nucleotide polymorphisms (SNPs) have been used by Miller et al. (2013). They indicate that North American subgenus *Vitis* species are monophyletic, i.e., descended from a common evolutionary ancestor (Fig. 2.2).

A. Rahemi et al., *Grape Rootstocks and Related Species*,
https://doi.org/10.1007/978-3-030-99407-5_2

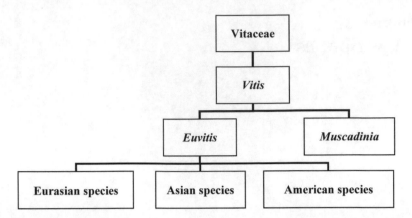

Fig. 2.1 Classification of *Vitis* genus [(Derived from Unwin (1996), and modified by Rahemi (2016)]

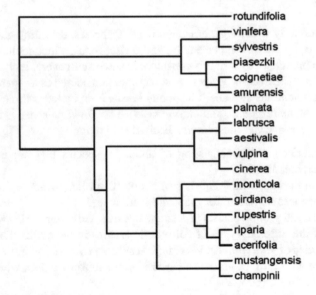

Fig. 2.2 *Vitis* species neighbor joining tree from a distance measure derived from hybridization intensities from the Vitis9KSNP array. Asian species (*V. amurensis*, *V. coignetiae*, and *V. piasezkii*), Eurasian species (*V. sylvestris*, and *V. vinifera*), and North American species (Euvitis subgenus: *V. acerifolia*, *V. aestivalis*, *V. champinii*, *V. cinerea*, *V. girdiana*, *V. labrusca*, *V. monticola*, *V. mustangensis*, *V. palmata*, *V. riparia*, and *V. rupestris*, and *V. vulpina*, and Muscadinia subgenus: *V. rotundifolia*) (Miller et al. 2013)

2.2   Grape Species Used in Rootstock Breeding Programs

The most important species that have been involved in the breeding of rootstocks are shown in Table 2.1. These species are either used directly as rootstocks or have been engaged in hybridization to produce the wide range of rootstocks that are currently in use.

Morphological and some viticultural details of each of these species follow in the sections below. The variation of the leaf blades of some *Vitis* species are shown in Fig. 2.3 and Fig. 2.4, and the characteristics of some species and their tolerance to abiotic and biotic stresses are provided in Table 2.2.

Table 2.1   The species that have been used in rootstock breeding programs (Rahemi 2016)

Scientific name	Subgenus	Groups
Vitis acerifolia Rafinesque	*Euvitis*	North American species
Vitis aestivalis Michaux	*Euvitis*	North American species
Vitis amurensis Ruprecht	*Euvitis*	Asian species
Vitis berlandieri Planchon	*Euvitis*	North American species
Vitis candicans Engelman ex Gray	*Euvitis*	North American species
Vitis caribaea De Candolle	*Euvitis*	North American species
Vitis × *champinii* Planchon	*Euvitis*	North American species
Vitis cinerea (Engelm.) Engelman ex Millardet	*Euvitis*	North American species
Vitis cordifolia Lamarck	*Euvitis*	North American species
Vitis labrusca Linnaeus	*Euvitis*	North American species
Vitis lincecumii Buckley	*Euvitis*	North American species
Vitis longii Prince	*Euvitis*	North American species
Vitis monticola Buckley	*Euvitis*	North American species
Vitis mustangensis Buckley	*Euvitis*	North American species
Vitis riparia Michaux	*Euvitis*	North American species
Vitis rotundifolia Michaux	*Muscadinia*	North American species
Vitis rufotomentosa Small	*Euvitis*	North American species
Vitis rupestris Scheele	*Euvitis*	North American species
Vitis simpsonii Munson	*Euvitis*	North American species
Vitis solonis (Planchon) Engelman ex Millardet	*Euvitis*	North American species
Vitis vinifera Linnaeus	*Euvitis*	Eurasian species
Vitis vulpina Linnaeus	*Euvitis*	North American species

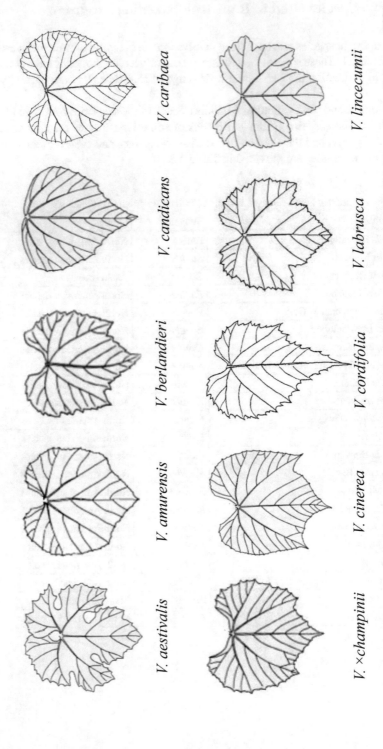

Fig. 2.3 Variation of the leaf blades in some wild grape species. *V. amurensis* Yamashita and Mochioka (2014); *V. labrusca*—Laguna Lumbreras (2004); *V. lincecumii* others—Bailey (1934)

Table 2.2 Characteristics of grape species and their tolerance to abiotic and biotic stresses

Species	Characteristics			Tolerance to abiotic stresses								Tolerance to biotic stresses							
	Vigorous	Ease of propagation	Root system	Cold	Heat	Drought	Wet	Salinity	Lime	Low pH	Phylloxera	Nematode	Crown gall	Pierce's disease	Downy mildew	Powdery mildew	Leaf roll virus	Fusariosis	Black rot
V. aestivalis	H	L		M/VH	H	H	M				14			H	H				
V. berlandieri	M	L	D	M	H	H	M		H		19			H	H	H			H
V. candicans	VH			M	H	H	M									H			
V. champinii	VH		D	M	H	H	H	H	H		15	VH		H					
V. cinerea						H	VH		H						H	H	H		
V. cordifolia	H			H	H	H	H				18	H			H				
V. labrusca	M	L		VH	M	M	M			H	5								
V. lincecumii	H			M	H	VH	H				14								
V. longii	VH			H	H	H	VH				14	H							
V. monticola	M			H	M	H	M				18								
V. riparia	H/L	H	S	VH	M	M	H			L	19		H		H	H	H	L	H
V. rotundifolia	H	H		M	H	H	M				20					H			
V. rupestris	VH	H	D	M	H	H	H			L	19			H	H	H			H
V. vinifera	M/VH	H		M	VH	H	M	H	H		1								

(Adapted/derived from Snyder 1937; Staudt 1997; Stafne and Carroll 2006)

2.2.1   Vitis acerifolia Rafinesque (Mapleleaf Grape)

It is reported that *V. longii* and *V. solonis* are part of *V. acerifolia* (Laguna Lumbreras 2004; Moore and Wen 2016). *Vitis solonis* was also reported to be a synonym for *V. longii* (Bailey 1934). *Vitis acerifolia* is a moderately high climbing vine with a large amount of lateral growth. Vines typically inhabit moist to slightly drier sites such as riverbanks and alluvial flood plains in woodlands. In more urban areas, they can also inhabit hedgerows and grow along fences. Young shoots and leaves have pubescence that is arachnoid (thin, soft, and entangled). After elongation, the shoot pubescents can become glabrous or may remain arachnoid. The bark on canes will persist for several years, then begin shedding in thin plates on older wood. Lenticels are absent or inconspicuous on shoots and canes. Leaves are often partially folded upward (conduplicate) with pubescence that is either arachnoid or glabrate. Leaves are often 3-lobed with sharply dentate edges, an overall cordate shape, and an acuminate apex. Clusters have short peduncles and typically have more than 25 small (512 mm), black, and glaucous berries that do not have obvious lenticels. Flowers appear from April to May, with fruit ripening from July to August (Moore 1991). *V. acerifolia* is considered the second most cold-hardy grape species in America after *V. riparia*. Importantly, *V. acerifolia* can continue growing during mild autumn weather and will harden off quickly once cold weather occurs. It does have phylloxera resistance, good root ability and is tolerant to Pierce's Disease (Table 2.2) (McGregor and Barkley 1986).

2.2.2   Vitis aestivalis Michaux (Summer Grape)

Vitis aestivalis is a high climbing vine generally found on well-drained sites. These include woodlands, thickets, scrubland, stabilized dunes, and much less often along riverbanks or floodplains. In urbanized areas, it can be found along fences and hedgerows. The current season's shoots are often tomentose or floccose but can occasionally be glabrous. Mature bark has inconspicuous or absent lenticels and exfoliates. Leaves have a cordiform to orbicular shape and can be unlobed, three-lobed, or even five-lobed (Fig. 2.3). Sinuses in lobed specimens can be rounded to acute. Leaf margins are crenate to dentate. The upper surface of the leaf can be glabrous to puberulent, while the lower surface can be arachnoid to floccose. The pubescence is commonly a rust color but maybe whitish. A cluster normally has more than 25 small to medium-sized (8–20 mm) black berries that do not show the presence of lenticels. Flowering occurs from April to June, with fruit ripening from July to September (Moore 1991). *V. aestivalis* is used as a wine grape in the central US. However, due to relatively low phylloxera resistance (Table 2.2), it has seldom been used in rootstock breeding (Stafne et al. 2015).

2.2.3   *Vitis amurensis Ruprecht (Amur Grape)*

Vitis amurensis is native to China, where it evolved tolerance to low temperature. For some selections, temperatures as low as − 40 °C can be survived. In addition, it shows high levels of resistance to many diseases, including grape white rot and anthracnose (Zhang et al. 2015). These traits have led Chinese breeders to use *V. amurensis* as a starting point for developing cold-resistant rootstocks (Shao-Hua 2001). Young shoots have sparse arachnoid tomentum. Leaves are entire with a broadly oval shape (Fig. 2.3), which starts off with sparse arachnoid tomentum that turns glabrescent. Leaf margins have irregular sharp serrations (Ren and Wen, 2007).

2.2.4   *Vitis berlandieri Planchon (Spanish Grape)*

Vitis berlandieri is a moderately climbing, stocky to stout vine. Young shoots start off with a gray cottony to tomentose pubescence, which they keep for much of the first year. Leaves are overall cordate shaped, entire, and commonly wider than long (Fig. 2.3). Margins have serrations of different sizes, which are broad or blunt. The upper surface of the leaf starts off with thin pubescence but becomes glabrous and dark green. The bottom of the leaf starts off cottony and keeps a level of short pubescence at maturity. The cluster starts at a long and stout peduncle. Berries are small (4–7 mm), globular, purple, or black, with a heavy bloom (Bailey 1934). Bloom is much later than other American species (late June), with fruit maturity around mid-October and leaf drop in late November (Galet 1979). *V. berlandieri* has lime tolerance, which is absent in most American species. It also has good phylloxera resistance, and some selections also have root-knot nematode resistance (Table 2.2). Unfortunately, it has low rooting and grafting ability (Walker 1991), making it more difficult to incorporate into breeding programs efforts successfully. Lime tolerance is an important trait in rootstock breeding as many of Europe's grape-growing regions are on limestone-based soils (Foëx 1902; Viala and Ravaz 1903). It has, therefore, found commercial success through some strategic hybridization with other species.

2.2.5   *Vitis candicans Engelmann*

Vitis candicans is a very vigorous and high climbing vine. Young growth is woolly or cobwebby with white or slightly tinted tomentum, which becomes floccose or ragged late in the season. Canes and other mature wood are strongly striated. The leaves (blades) themselves appear thicker than other American species. The leaves are triangular to cordate-ovate in shape and overall small (8–14 cm) (Fig. 2.3). The leaves can be entire or lobed, with margins sinuate or obscurely shallow serrated.

Clusters have a short peduncle and have 12 to 20 large (12–22 mm), black to reddish to whitish berries with thick, tough skin, which ripens in August (Bailey 1934).

2.2.6 Vitis caribaea De Candolle (Syn. Vitis tiliifolia Humboldt and Bonpland Ex Schultes)

Vitis caribaea (or V. tiliifolia) is found in Mexico, the Caribbean, Central America, and parts of Western South America (USDA 2019). Canes are woody and obtain a diameter of up to 8 mm with dark brown and rough bark. Leaves are entire with a broadly ovate or trilobate shape with serrated margins (Fig. 2.3). The upper surface of leaves has a puberulent tomentum, while the lower surface has tomentose tomentum. Berries are small (5–10 mm) and violet in color (Acevedo-Rodriguez 2005). V. caribaea does have tolerance to Pierce's Disease (Table 2.2); however, like many other grapes from the Southern United States, it still supports the pathogen (Shi et al. 2007). Hybrids of V. caribaea, were created in Brazil for tropical regions, where it has superior performance in terms of productivity (Dalbó and Feldberg 2016).

2.2.7 Vitis × champinii Planchon (Champin's Grape)

Vitis × champinii is not a true grape species but a wild hybrid of V. mustangensis and V. rupestris. V. × champinii is a high climbing and vigorous vine. Young growth has arachnoid pubescence that becomes glabrate with maturity. Leaves are cordate in shape that is either 3-shouldered or occasionally 3-lobbed with margins crenate to slightly serrate (Fig. 2.3). The upper surface of the leaf is typically glabrous or occasionally lustrous, with the lower surface thinly arachnoid to glabrate. Clusters usually have less than 25 small (12 mm) black berries. Flowering occurs from April to May, with fruit ripening from July to August (Moore 1991). V. × champinii grows in wooded areas with well-drained calcareous soil low in organic matter. It does have good root-knot and variable dagger nematode resistance. As with most American species, it does have phylloxera resistance (Table 2.2), but its durability is not well understood (Walker 1991).

2.2.8 Vitis cinerea Engelmann (Winter Grape, Grey Bark Grape)

Vitis cinerea is a high climbing vine. Current season shoots are covered in dense trichomes and/or dense arachnoid pubescence. Leaves also have dense trichomes and/or dense white arachnoid pubescence, especially on the bottom surface. Leaves are cordiform in shape, unlobed to 3-shouldered, or on occasion 3-lobed with crenate to dentate margins (Fig. 2.3). Clusters normally contain more than 25 very small (4– 8 mm) black berries. Flowering occurs from May to June, with fruit ripening from July to October. *V. cinerea* usually grows in floodplains, lowland woods, along stream banks, and pond margins. In urban areas, it can be found along fence rows (Moore 1991).

2.2.9 Vitis cordifolia Michaux (Frost Grape)

Vitis cordifolia is a vigorous and tall vine with a very thick trunk. Young growth is vigorous and can have a grayish to tawny pubescence but becomes glabrous with maturity. Leaves are cordate-ovate or triangle in shape (Fig. 2.3), firm-textured, and overall small (in comparison to *V. labrusca* or *V. vinifera*). Leaves are normally unlobed but can be shouldered with a simple toothed margin. The upper surface of the leaves is glabrous and glossy, and the lower surface is also glabrous with tufts of hairs in the vein intersections with a light green color. Clusters are slender and open and frequently have a forked base. Berries are small (3–9 mm), black, and mature in September and October (Bailey 1934).

2.2.10 Vitis labrusca Linnaeus (Fox Grape)

Vitis labrusca is a high climbing vine. It grows in a wide range of habitats, including upland well-drained sites, lowland poorly drained sites, to intermittently flooded bottomlands. Young growth is normally coated in dense tomentum and occasionally spinose pubescence. Leaves are cordiform in shape normally with three shoulders and crenate to crenate-dentate margins (Fig. 2.3). The upper surface of the leaves is normally glabrous but can be slightly pubescent and dull in color. The lower surface is covered in whitish to yellowish dense arachnoid tomentum. The tomentum on the lower surface is often thick enough to obscure the lower leaf surface, but not the veins. Clusters generally have less than 25 medium-sized (greater than 12 mm) black berries. Flowering occurs from May to June, and fruit ripening from September to October (Moore 1991).

2.2.11 Vitis lincecumii Buckley (Post Oak Grape)

Vitis lincecumii is a shorter climbing vine that will form a bush when no objects are available for support. Young shoots are covered in a dense whitish or lightly brown tomentose. Leaves are cordate-ovate in shape, occasionally entire or slightly shouldered, but are often 3-lobed or 5-lobed (Fig. 2.3). The sinuses on lobed leaves are often open with a round shape. Leaf margins are coarsely shallow toothed. Leaves are heavily pubescent at first but do thin out with age. Even leaf petioles will have a tomentose pubescence. The pubescence on both can be gray or tawny. Clusters are fairly short (8–10 cm) with large (10–25 mm) black or dark purple berries. Fruit generally matures from July to August (Bailey 1934).

2.2.12 Vitis longii Prince (Bush Grape)

Vitis longii is a short and bushier vine that often only climbs over rocks and shrubs. Young growth is tomentose to floccose with a whitish or grayish color. Leaves are generally cordate-ovate to almost circular in shape, shallowly lobed or shouldered, with coarse angular and short serrations on the margin (Fig. 2.4). Clusters are short (3–7 cm) with small (8–12 mm) black berries. The fruit is generally ripe in July or August (Bailey 1934). *V. longii* can grow in limestone areas with sandy soils that are leached with little organic matter. They can grow in areas that are often dry but punctuated with heavy rains. This species does have good root-knot nematode, and some dagger nematode resistance, along with moderate phylloxera resistance. It also roots and grafts well (Walker 1991). *V. longii* has been grouped, along with *V. solonis*, and with *V. acerifolia* in some taxonomic work (Laguna Lumbreras 2004).

2.2.13 Vitis monticola Buckley (Sweet Mountain Grape)

Vitis monticola is a moderate to high climbing vine (Moore and Wen 2016). Young growth has slightly to moderately arachnoid pubescence. Leaves are cordiform in shape, either 3-shouldered or shallowly 3-lobed, with irregularly dentate-serrate teeth on the margins (Fig. 2.4). The upper leaf surface is normally glabrous, and the lower surface is glabrous to sparsely hirtellus. Clusters typically have less than 25 small (8–10 mm) black berries with lenticels. Flowering normally happens in May, with fruit maturing from July to August (Moore 1991).

2.2.14 Vitis mustangensis Buckley (Mustang Grape)

Vitis mustangensis is a vigorous, high climbing vine. Generally found in woodlands, woodland borders, thickets, and lowland woods. Young growth has a densely tomentose tomentum, which becomes thinly tomentose with age. Leaves are broadly cordate to reniform in shape, concavely folded, with shallow, broad scalloped, obtuse teeth on the margin. Leaves are normally unlobed, but some 3-shouldered and even fewer with 3 to 5 lobes can be found. The upper leaf surface is glabrous, while the lower surface is densely covered with white to rusty tomentum thick enough to obscure

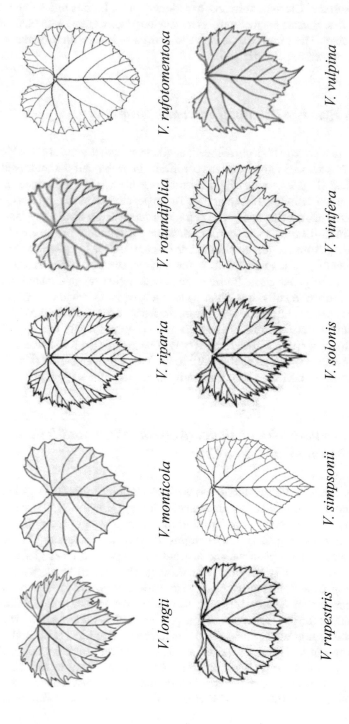

Fig. 2.4 Variation of the leaf blades in some wild grape species. *V. riparia, V. rupestris, V. solonis, V. vinifera,* and *V. vulpine*—Laguna Lumbreras (2004); others—Bailey (1934)

the lower leaf surface. Clusters normally have less than 25 berries and sometimes less than 12 berries. Berries are medium in size (greater than 12 mm) with black and rarely dark red color. Their flowering is in late May to early June, with fruit maturing from August to September (Moore 1991).

2.2.15   Vitis riparia Michaux (Riverbank Grape)

Vitis riparia is a moderate to high climbing vine. Young growth is glabrous to slightly hirtellous in pubescence. Leaf-blades are cordiform in shape and 3-shouldered to shallowly 3-lobed (Fig. 2.4) with slight to moderate hirtellous pubescence. Leaf margins have sharply dentate serrate margins. The upper surface of the leaf is usually glabrous and often has a yellowish-green color. Clusters usually have more than 25 small to medium (8–12 mm) black berries. Flowering is typically in April to June, with fruit ripening from August to September (Moore 1991). V. riparia normally grows in areas with continual access to water, such as streams and riverbanks or fence lines next to irrigated grass. It prefers soils with higher organic matter and is not adapted to limestone soils. It has the widest geographical range of any North American grape species and can be found from the Rocky Mountains to the Atlantic Ocean and from Canada to Texas. V. riparia is easy to propagate and generally has good phylloxera resistance. However, it only contains limited resistance to both dagger and root-knot nematodes (Table 2.2). It also goes dormant earlier than other species and appears to hasten or induce maturity in grafted scions (Walker 1991).

2.2.16   Vitis rotundifolia Michaux (Muscadinia rotundifolia Michaux) (Muscadine Grape)

Vitis rotundifolia (Muscadinia rotundifolia) is a high climbing vine. Unlike all other Vitis, tendrils on V. rotundifolia are unbranched. Young growth usually has thin, grayish arachnoid pubescence. Leaves are cordiform to nearly reniform in shape and only rarely have lobes (Fig. 2.4). Leaf margins are crenate to dentate. The upper surface of mature leaves is glabrous and lustrous, with lower surfaces glabrous or pubescent. Clusters normally have less than 25 large (8–25 mm) black, purple, or occasionally bronze berries. Flowering occurs in April or May, with berries ripening in late July to September (Moore 1991). In the year-old canes of V. rotundifolia, the diaphragm is absent (the cane contains a large pith, and pith is interrupted by woody tissue at the nodes; this woody tissue is the diaphragm) (Hedrick et al. 1908). V. rotundifolia grows in soils that are acidic and have relatively high organic matter in locations that are not overly wet nor overly dry. It is not easy to root (Goode et al. 1982) or graft (Bouquet and Hevin 1978; Bouquet 1980), but it does appear to be fully resistant to most soil pests of V. vinifera (Walker 1991). The chromosome number in muscadine ($2n = 40$) is different from *V. vinifera* ($2n = 38$) (Lewter et al. 2019), which sometimes can cause difficulty in hybridization. Attempts to produce *V. rotundifolia* × *V. vinifera* hybrids that combine good fruit quality and disease resistance have been reported, but there was limited success when *V. vinifera* was used as the seed parent (Lu et al. 2000).

2.2.17   *Vitis rufotomentosa Small*

Sometimes *Vitis rufotomentosa* is considered as a subspecies of *V. aestivalis* (USDA 2019). *V. rufotomentosa* has young growth with dense rusty tomentum. Leaves are broadly cordate-ovate to ovate, have entire or obscure lobes on exposed areas, and can be 3–5 lobed on ground shoots with deep wounded sinuses (Fig. 2.4). Margins are shallowly sinuate-toothed. The upper leaf surface is dull with thinly webby tomentum, and the lower surface has loose webby tomentum. Clusters are short (<12 cm) with small (5–6 mm) black berries (Bailey 1934). *V. rufotomentosa* has shown to have good resistance to *Xiphinema index* (Table 2.2) (Harris 1983).

2.2.18   *Vitis rupestris Scheele (Sand Grape)*

Vitis rupestris is a low climbing, typically shrubby vine that is heavily branched. Young shoots are glabrous to slightly hirtellous pubescent. Leaves have a reniform shape and are conduplicate folded (Fig. 2.4). The leaves are often 3-shouldered or rarely have three shallow lobes. Leaf margins are dentate-serrate, with upper surface glabrous and lustrous, with lower surface typically green and glabrous, but occasionally sparsely hirtellous. Clusters normally have less than 25 small (8–12 mm) black berries. Flowering occurs from April to May, with fruit ripening from August to September (Moore 1991). *V. rupestris* grows in rocky stream beds that are generally not very fertile from excessive leaching. It does not grow well on limestone-rich soils but has good propagation and grafting characteristics. It has good phylloxera resistance but is more susceptible to nematodes than some other species (Walker 1991). It is reported that *V. rupestris* is susceptible to chlorosis (Ollat et al. 2016). However, Branas (1974) reported that it is slightly tolerant (Table 2.2). There is evidence to suggest its use as a rootstock results in chlorosis in the scion (Ollat et al. 2016).

2.2.19   *Vitis simpsonii Munson (Simpson's Grape)*

Vitis simpsonii has been previously considered as two different taxa: (1) under *Vitis cinerea* var. Floridana of series Cinerascentes, and (2) under *Vitis aestivalis* var.

Smalliana of series Aestivalis (Comeaux and Fantz 1987). Modern genetic studies have shown that all tested samples previously assigned to *V. simpsonii* were part of *V. aestivalis* (Wan et al. 2013). Leaf-blades are densely tomentose with rusty or tawny hairs (Fig. 2.4).

2.2.20   Vitis solonis (Planch.) Engelman Ex Millardet

V. solonis has leaves that are entire, with curved sharp teeth on the margins and bristly pubescence on the underside (Fig. 2.4). Clusters are small with small black berries. It grows on wet, clay-based soils and includes some salt tolerance but is not drought tolerant (Vinograd 2019). Padgett-Johnson et al. (2003) also reported that *V. solonis* was not drought tolerant. Husmann (1908) reported that *V. solonis* varieties have fair growth on sandy or lime soils and root easily.

2.2.21   Vitis vinifera Linnaeus

Vitis vinifera is a moderately high climbing grapevine with a sprawling habit and limited lateral growth. Growing shoots can be pubescent to glabrescent, with growing tips not enveloped in unfolding leaves. Leaves are cordate-ovate to cordate-orbiculate in shape and can be entire to deeply 5-lobed (Fig. 2.4). The under surface of the leaf can be sparingly pubescent to glabrate. Clusters have small to large (825 mm) reddish-purple to nearly black berries with skin adhering to the pulp (Moore and Wen 2016). Leaf margins have large, irregular, and sharp teeth (Ren and Wen 2007). *V. vinifera* grows in wooded areas and streambanks to dry hillsides and has excellent tolerance to both drought and calcareous soils. It is excellent at rooting and grafting. It has some resistance to root-knot nematodes and mild tolerance to fanleaf virus (Table 2.2). It is fully susceptible to root-feeding phylloxera and appears to pass a dominant gene for phylloxera root susceptibility to offspring (Walker 1991).

2.2.22   Vitis vulpina Linnaeus

Vitis vulpina is a high climbing vine. Young growth is glabrous to sparsely arachnoid in pubescence. Leaves are cordiform in shape (Fig. 2.4) and can be 3-shouldered to shallowly 3-lobed with deep lobes on ground shoots. Leaf margins are irregularly dentate-serrate. The upper surface of leaves is glabrous or occasionally has sparse hirtellous pubescence. The lower leaf surface is typically green with short, straight hirtellous pubescence. Clusters usually have more than 25 small (8–12 mm) black berries. Flowering occurs in May, with fruit ripening from July to August. In natural habitats, *V. vulpina* typically grows in well-drained woodlands or woodland borders but uncommonly is found in floodplains or lowland woods. In suburban areas, it can also be found along fences and hedgerows (Moore 1991). *Vitis vulpina* showed high resistance to downy mildew (*Plasmopara viticola*) and fusariosis (*Fusarium oxysporum* f.sp. herbomontis) (Table 2.2) (Gu 2010). Segregation of resistance to root-knot nematodes was reported in a *V. vulpina* hybrid population (Cousins and Lauver 2003). *V. vulpina* could be used for rootstock breeding (Stafne and Carroll 2006).

References

Acevedo-Rodriguez P (2005) Vines and climbing plants of Puerto Rico and the Virgin Islands. Contrib US Natl Herbarium 51:1–483

Bailey L (1934) The species of grapes peculiar to North America. Gentes Herbarium, vol 3. Fasc. I, Ithaca, New York, pp 149–244

Bouquet A, Hevin M (1978) Green-grafting between Muscadine grape (*Vitis rotundifolia* Michx) and bunch grapes (*Euvitis sp.*) as a tool for physiological and pathological investigations. Vitis 17:134–138

Branas J (1974) Viticulture. Imprimerie Dèhan, Montpellier, p 990

Christenhusz MJM, Byng JW (2016) The number of known plant species in the world and its annual increase. Phytotaxa. Magnolia Press 261(3):201–217. https://doi.org/10.11646/phytotaxa.261.3.1

Comeaux BL, Fantz PR (1987) Nomenclatural clarification of the name *Vitis simpsonii* Munson (Vitaceae). SIDA, Contrib Botany 12(2):279–286

Cousins P, Lauver M (2003) Segregation of resistance to root-knot nematodes in a *Vitis vulpina* hybrid population. Acta Hortic 623:313–318. https://doi.org/10.17660/ActaHortic.2003.623.35

Dalbó MA, Feldberg N (2016) Agronomic behavior of grape rootstocks resistant to young vine decline in Santa Catarina State, Brazil, BIO Web of Conferences 7, 01017 (2016), 39th world congress of Vine and Wine, 1–3. https://doi.org/10.1051/bioconf/20160701017

Foëx G (1902) Manual of modern viticulture: reconstitution with American vines. In: Dubois R, Wilkinson WP, Brain RS (eds) Translated from the sixth French edition. Government Printer, Melbourne, Australia

Galet P (1979) A practical ampelography, Translated and adapted by Lucie T. Morton. Ithaca, NY: Comstock Cornell University Press

Goode Jr DK, Krewer GW, Lane RP, Daniell JW (1982) Rooting studies of dormant muscadine grape cuttings. HortScience 17:644–645

Gu S (2010) Effect of rootstocks on grapevines. Kentucky State University: http://www.pawpaw.kysu.edu/viticulture/Information/Rootstock

Harris AR (1983) Resistance of some *Vitis* rootstocks to *Xiphinema index*. J Nematol 15:405–409

Hedrick UP, Booth NO, Taylor OM, Wellington R, Dorsey MJ (1908) The grapes of New York. New York (State). Dept. of Agriculture. Agricultural Experiment Station. Albany: J.B. Lyon Co., state printers. p 788

Husmann GC (1908) The adaptability of grapevines to different vineyard soils. In: Proceeding of thirty-third fruit growers convention, pp 238–250

Laguna Lumbreras E (2004) American and hybrid grapevines (*Vitis spp.*): a new concept of invasive plants to Europe. Proceedings of the 4th European conference on the conservation of the wild plants, a workshop on the implementation of the global strategy for plant conservation in Europe, Valencia, Spain. https://www.researchgate.net/figure/Fig-n-1-Representation-of-leaves-of-several-pure-species-of-Vitis-L-most-of-them_fig1_267960737

Lewter J, Worthington ML, Clark JR, Varanasi AV, Nelson L, Owens CL, Conner P, Gunawan G (2019) High-density linkage maps and loci for berry color and flower sex in muscadine grape

(*Vitis rotundifolia*). Theor Appl Genet 132(5):1571–1585. https://doi.org/10.1007/s00122-019-03302-7

Lu J, Schell L, Ramming DW (2000) Interspecific hybridization between *Vitis rotundifolia* and *Vitis vinifera* and evaluation of the hybrids. Acta Hortic 528:479–486. https://doi.org/10.17660/Act aHortic.2000.528.69

McGregor RL, Barkley TM (1986) Flora of the great plains. University Press of Kansas, The great plains flora association, p 1392

Miller AJ, Matasci N, Schwaninger H, Aradhya MK, Prins B, Zhong GY, Simon C, Buckler ES, Myles S (2013) Vitis phylogenomics: hybridization intensities from a SNP array outperform genotype calls. PLoS ONE 8(11):e78680. https://doi.org/10.1371/journal.pone.0078680.g001

Moore MO (1991) Classification and systematics of eastern North American *Vitis* L. (Vitaceae). North of Mexico. Sida 14:339–367

Moore MO, Wen J (2016) Vitaceae. In: Flora of North America, North of Mexico, Flora of North America editorial committee, vol 12. New York, NY; Oxford: Oxford University Press

OIV, International Organisation of Vine and Wine (2017) Distribution of the world's grapevine varieties, OIV Publication, p 54. ISBN: 979–10–91799–89–8. http://www.oiv.int/public/medias/5888/en-distribution-of-the-worlds-grapevine-varieties.pdf

Ollat N, Peccoux A, Papura D, Esmenjaud D, Marguerit E, Tandonnet JP, Bordenave L, Cookson SJ, Barrieu F, Rossdeutsch L, Lecourt J, Lauvergeat V, Vivin P, Bert PF, Delrot S (2016) Rootstocks as a component of adaptation to environment. In: Gerós H, Chaves MM, Gil HM, Delrot S (eds). Grapevine in a changing environment: a molecular and ecophysiological perspective. https://doi.org/10.1002/9781118735985.ch4

Padgett-Johnson M, Williams LE, Walker MA (2003) Vine water relations, gas exchange, and vegetative growth of seventeen *Vitis* species grown under irrigated and non-irrigated conditions in California. J Am Soc Hortic Sci 128(2):269–276

Rahemi A (2016) Breeding of grapevine rootstocks for adaptability to the environmental stresses, Third national conference of grape and raisin conducted by research institute of grape and raisin (RIGR), 27–29 September 2016. Malayer University, Malayer, Iran

Ren H, Wen J (2007) Vitis. In: Wu Z-Y, Hong D-Y, Raven PH (eds) Flora of China 12. Science Press, Beijing and Missouri Botanical Garden Press, St. Louis, pp 210–222

Shao-Hua L (2001) Grape production in China. In: Papademetriou MK, Dent FJ (eds) Grape production in the Asia-Pacific region. FAO publication, Expert consultation on "Viticulture (grape production) in Asia and the Pacific", held in Bangkok, Thailand, 2–4 May 2000

Shi XY, Dumenyo CK, Hernandez-Martinez R, Azad H, Cooksey DA (2007) Characterization of regulatory pathways in *Xylella fastidiosa*: genes and phenotypes controlled by *algU*. Appl Environ Microbiol 73:6748–6756

Snyder E (1937) Grape development and improvement. In: Yearbook of agriculture 1937. United States, Department of Agriculture, 75th Congress, 1st Session House Document vol 28, pp 631–664

Stafne ET, Carroll B (2006) Rootstocks for grape production. Oklahoma Cooperative Extension Service, HLA-6253, p 4

Stafne ET, Sleezer SM, Clark JR (2015) Grapevine breeding in the Southern United States. In: Reynolds A (ed) Grapevine breeding programs for the wine industry. Woodhead Publishing, pp 379–410

Staudt G (1997) Evaluation of resistance to grapevine powdery mildew (*Uncinula necator*) in accessions of *Vitis* species. Vitis 36:151–154

The Plant List (2013) Version 1.1. Published on the Internet http://www.theplantlist.org/. Accessed August 2019

Unwin T (1996) Wine and the vine, an historical geography of viticulture and the wine trade. Rouledge, Taylor & Francis Group, London and New York. p 436

USDA, United States Department of Agriculture (2019) Natural resources conservation service. https://plants.usda.gov/java/nameSearch?keywordquery=Vitis. Accessed Dec 2019

Viala P, Ravaz L (1903) American vines: their adaptation, culture, grafting and propagation. R Dubois and E.H. Twight, translators. Freygang-Leary Co., San Francisco. p 299

Vinograd (2019) Solonis. http://vinograd.info/spravka/slovar/solonis-2.html. Accessed 28 Sep 2019

Walker MA (1991) Characteristics of *Vitis* species commonly used in rootstock breeding. ASEV Publication, Davis, CA, Alternative Rootstock Update, pp 1–7

Wan Y, Schwaninger HR, Baldo AM, Labate JA, Zhong G, Simon CJ (2013) A phylogenetic analysis of the grape genus (*Vitis* L.) reveals broad reticulation and concurrent diversification during neogene and quaternary climate change. BMC Evol Biol 13: 141. http://www.biomedcen tral.com/1471–2148/13/141

Yamashita H, Mochioka R (2014) Wild grape germplasms in Japan. Adv Hort Sci 28(4):214–224. https://doi.org/10.1400/230471

Zhang QT, Fan ST, Lu WP, Song RG, Yang YM, Ai J (2015) Breeding progress of Amur grape (*Vitis amurensis* Purr.) in China. Acta Hortic 1082:33–36. https://doi.org/10.17660/ActaHortic. 2015.1082.1

Chapter 3
Grape Rootstocks Breeding

3.1 Breeding Programs in Different Countries

The main priorities of grape rootstock breeding programs in each country are based on resistance to local pests and desired adaptions to local climate pressures and conditions.

The Geisenheim Research Institute in Germany has traditionally focused on the selection of resistance to phylloxera based on the hypersensitive response obtained by *V. cinerea* 'Arnold' (Dry 2005). Geisenheim also attempts on rootstock and scion compatibility mechanism, and adaptation of rootstocks to different soil conditions such as drought and lime, and study on rootstocks effects on crop quality (Whiting 2012).

In the United States, Cornell University has focused their priority on nematode resistance and providing plant material with ongoing protection against phylloxera. Phylloxera tolerance or resistance is a necessity for all their elite selections, achieved by using *V. cinerea*, *V. cordifolia*, *V. longii*, and *V. solonis* (Cousins et al. 2007).

The National Institute of Agronomic Research (INRA) in France has focused their breeding program on finding resistance to the vector of grapevine fanleaf virus (GFLV) and dagger nematode (*Xiphinema index*) are using *Muscadinia rotundifolia* in their breeding program. They are also working on evaluating rootstocks from other countries for drought tolerance (Whiting 2012).

Similar to the INRA, the University of California (Davis, USA) is also using *Muscadinia rotundifolia* for resistance to the vector of GFLV, dagger nematode (*X. index*), and virulent populations of root-knot nematode (RKN). They are also working on resistance to drought and salinity by using *V. acerifolia, V. arizonica, V. berlandieri, V. doaniana,* and *V. girdiana* (Heinitz et al. 2015).

© The Author(s), under exclusive license to Springer Nature Switzerland AG 2022
A. Rahemi et al., *Grape Rootstocks and Related Species*,
https://doi.org/10.1007/978-3-030-99407-5_3

In Australia, grape rootstock breeding is targeted to address particular issues such as resistance to phylloxera and nematodes, graft incompatibility, absorption of potassium from the soil, salinity tolerance, chlorosis, pH of the soil, drought tolerance, root system, nutrition, vegetative and reproductive growth, bud break and bud fruiting, crop quantity and quality, propagation, disease, and viral issues, germplasm, industrial source blocks, and preparation of rootstock information. For example, at the CSIRO (Commonwealth Scientific and Industrial Research Organisation) in Merbein, Australia, they have focused on rootstocks with phylloxera and nematode resistance, reduced potassium uptake to address high juice pH issues, and the restricting uptake of the ions, such as sodium and chloride, ease of grafting, and propagation, vigor potential, water use efficiency, and wine quality induced by rootstocks (Whiting 2012).

In California Polytechnic State University the researchers are working on root system architecture and understanding the role of rootstocks on change to key phenological events expressed by the scions (Riffle et al. 2021).

There are many other notable advancements to breeding efforts and plant material availability as well that are recognizable. Those listed here are simply a brief list of some of the larger, present-day programs. For more information, consult The Australian Grape and Wine Authority (AGWA)(Australia), the Research Institute for Viticulture and Enology (Hungary), the University of Milan (Italy), Centro di Ricerca per la Viticoltura (CRA-VIT) (Italy), the IFV (French Institute for Vines and Wines), the Instituto Agronômico de Campinas (IAC) (Brazil) and the Simcoe Research Station at the University of Guelph (Canada) (Rahemi 2016).

3.2 Rootstocks Evaluation

Before releasing rootstocks, breeders must assess them. This is done through four different activities designed to evaluate performance: (1) vineyard production (individual yield and growth, disease incidence, scion compatibility). (2) nursery production (vine production, transplant evaluation). (3) laboratory assessment (molecular markers, tissue culture). (4) greenhouse trials. A grape rootstock breeding and selection protocol flowchart has been provided by Rahemi (2016) (Fig. 3.1).

Rootstock trials can be extremely demanding on resources and are long-term in nature. A checklist was developed by Morton (1995) to record essential information applicable to grape rootstock field trials. It included the following: (1) Experimental

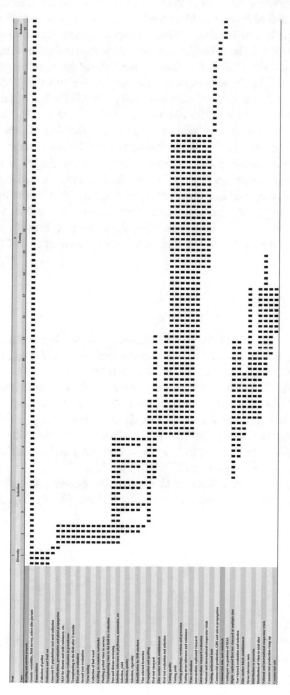

Fig. 3.1 Grape rootstock breeding and selection protocol (Rahemi 2016)

objectives and estimated duration. (2) Plant material. Reasons for selecting specific varieties of rootstocks and scion in the experiment, source of rootstock and scion, method of grafting, plant material condition and planting date for each graft combination, growth obtained at the end of growing season before collecting crop data, and note whether there is a significant difference between the plants within a graft combination group or not. (3) Location conditions. The reasons for choosing a specific site, climate, soil structure and chemistry, diseases, and pests. (4) Design of the vineyard. The reason for choosing the specific vineyard design, inter and intra-row spacing, trellising and pruning protocol, method of irrigation, replication of graft combinations, plan map of the vineyard and use the permanent labels to show each graft combination for each vine, and should not be allowed replanting after the second growing season. (5) Data from harvesting. Harvest time (at the specified sugar content or on the same date), crop yield, sugar content, total acidity (TA), minimum pH, also extract color and tannins etc., fruit condition (damage, and etc.), and weather conditions at the harvesting time. (6) Data collecting at postharvest. Pruning weight and measuring cane uniformity, the starch content of canes, and diameter of the trunk.

3.3   Selection Protocols

It takes several years to develop and release a new rootstock. For example, at INRA, they evaluate rootstocks at least for 10 years after selections were made. In 2010 they released 'Nemadex Alain Bouquet', which was bred in 1987.

The M series (M1, M2, M3, and M4) are new rootstocks that were introduced by the University of Milan (UniMI) over a period of 27 years. These rootstocks show tolerance to water and salt stresses and ferric chlorosis (Fig. 3.2).

The GRN rootstocks (GNR1, GNR2. GNR3. GNR4, and GNR5) have also been released by UC Davis over 28 years from designed breeding and a selection program on resistance to nematodes (Ferris et al. 2012) (Fig. 3.3).

A world grape rootstocks pedigree that shows each rootstock breeding pathway and their interrelationships have been provided by Rahemi (2016) (Fig. 3.4).

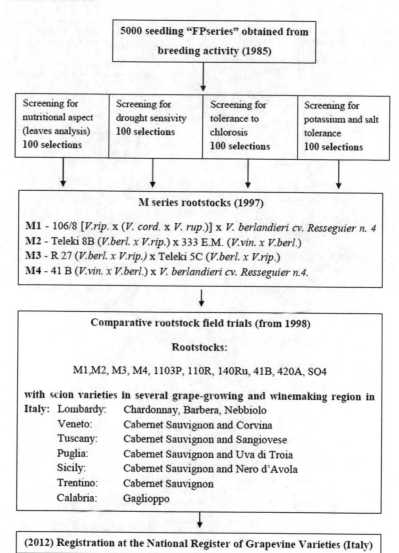

Fig. 3.2 Screening processes involved in the development of the M series of rootstocks at the University of Milan, Italy (Grossi 2015)

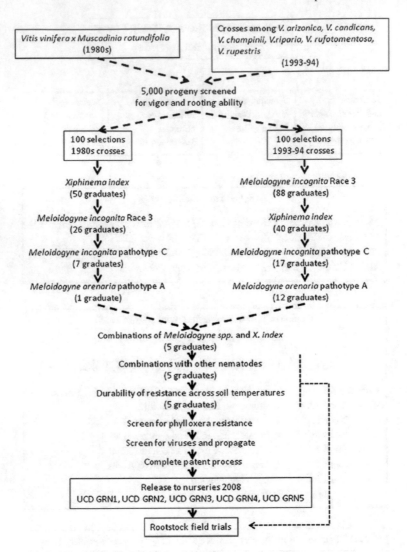

Fig. 3.3 Screening processes involved in the development of the GRN series of rootstocks at the UC Davis, U.S. (Ferris et al. 2012)

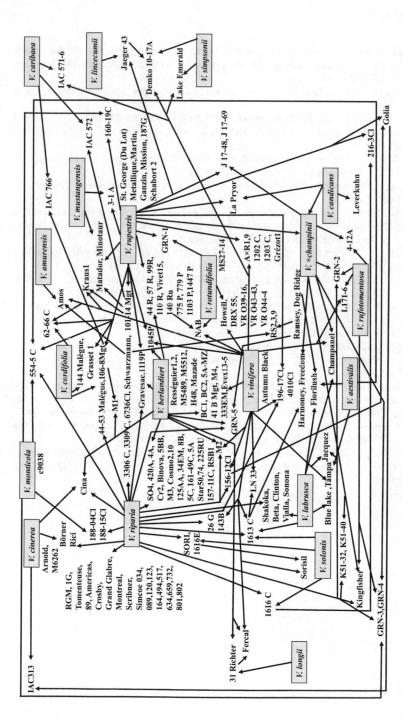

Fig. 3.4 World grape rootstocks pedigree. *Sources* Candolfi-Vasconcelos et al. (1995), Shaffer et al. (2004), Corso and Bonghi (2014), and Schmid et al. (2015). Modified extensively by Rahemi (2016)

References

Candolfi-Vasconcelos MC, Fisher GC, Hellman H, Watson B, Price S, Connelly AE, Stonerod P, Strik BC (1995) Phylloxera: strategies for management in Oregon's vineyards. Or State Univ Extension Catalog, EC 1463:24p

Corso M, Bonghi C (2014) Grapevine rootstock effects on abiotic stress tolerance. Plant Sci Today 1(3):108–113

Cousins P, Johnston D, Switras-Meyer S, Boyden L, Vidmar J, Meyer C (2007) USDA ARS research in grape rootstock breeding and genetics. Acta Hortic 733:51–58. https://doi.org/10.17660/Act aHortic.2007.733.5

Dry N (2005) Final report to grape and wine research and development corporation, phylloxera and grape industry board of South Australia, PGI 05/01 p 22

Ferris H, Zheng L, Walker MA (2012) Resistance of grape rootstocks to plant-parasitic nematodes. J Nematol 44(4):377–386

Grossi D (2015) Eco-physiological characterization of new grapevine rootstocks under drought stress, DiSAA—department of agricultural and environmental sciences, production, landscape, agroenergy of the University of Milan. PhD. dissertation p 101

Heinitz CC, Fort K, Walker MA (2015) Developing drought and salt resistant grape rootstocks. Acta Hortic 1082:305–312. https://doi.org/10.17660/ActaHortic.2015.1082.42

Morton L (1995) Recommendations for the study of grape rootstocks in New Zealand. N Z J Crop Hortic Sci 23(4):403–405. https://doi.org/10.1080/01140671.1995.9513916

Rahemi A (2016) Breeding of grapevine rootstocks for adaptability to the environmental stresses, third national conference of grape and raisin conducted by research institute of grape and raisin (RIGR), 27–29 September 2016. Malayer University, Malayer, Iran

Riffle V, Palmer N, Casassa LF, Dodson Peterson JC (2021) The effect of grapevine age (*Vitis vinifera* L. cv. Zinfandel) on phenology and gas exchange parameters over consecutive growing seasons. Plants 10(2):311

Schmid J, Manty F, Rühl EH (2015) Optimisation of sustainability of grapevine varieties by selecting rootstock varieties under different environmental conditions and creating new rootstock varieties. Retrieved 5 Oct 2019. http://www.viticulturasustentavel.com/pdfs/Prof_Joachim_Schmid.pdf

Shaffer R, Sampaio TL, Pinkerton J, Vasconcelos MC (2004) Grapevine rootstocks for Oregon vineyards, OSU extension Catalog. https://catalog.extension.oregonstate.edu

Whiting JR (2012) Rootstock breeding and associated R&D in the viticulture and wine industry, Australian government, grape and wine research and development corporation (GWRDC). https://gwrdc.com.au

Chapter 4
Grape Scion and Rootstock

4.1 The Importance of Grape Rootstocks

Grape rootstocks are used for adaptation to soil pests and soil conditions. Soil conditions can include adaptation to specific types of soil, soil texture, soil depth, soil chemistry (pH, salinity, lime content), and soil nutrient availability (fertility). They are also used to overcome water problems, such as drought conditions or excess water (inadequate drainage). Rootstocks are also used to be compatible with vineyard and winemaking objectives. This can include their ability to affect vine growth, the timing of fruit maturity, and fruit quality (Christensen 2003). The difference in disease incidence in rootstocks is very important and could be helpful in controlling the disease (Ferreira 1985).

The main criteria identified for selecting rootstocks include phylloxera resistance, resistance to nematodes, compatibility with high or low pH soils, adaptability to saline soils, adaptability to wet or poorly drained soils, and adaptation to drought (Reynolds and Wardle 2001). At some production sites, adaptability to winter conditions will also be critical (Rahemi 2016). Consequently, in the Cornell University grape rootstock breeding program, the selection is focused on providing increased resistance to soil-borne pests, such as phylloxera or nematodes; combatting replant effects (primarily high initial phylloxera populations, but also the likely impacts of nematodes and crown gall bacteria); providing increased lime tolerance (Ca); providing a larger root system to improve vine drought tolerance; providing cold-tolerant roots and trunk; reducing the chance of virus transmission by nematodes; and conferring tolerance to low soil pH (Pool et al. 2005). The main traits evaluated for selecting a new rootstock in an Australian breeding program are phylloxera tolerance; root-knot nematode tolerance (egg masses/dried root weight (g)); estimated breeding value; K uptake; inheritance of rooting ability, root system architecture; and graft union compatibility (Clingeleffer and Davis 2012). Further, enough plant material (lignified cane) with the proper diameter needs to be produced for propagation. There also needs to be a low potential to produce suckers, as this prevents the long canes required during grafting. Some special characteristics, such as the color of leaves and canes, and the shape of the leaves, are useful for distinguishing a rootstock from other cultivars in the nursery (Rahemi 2016).

Appropriate statistical analysis and information on related plants and families (parents, siblings, progeny) can be used to improve the estimates of the usefulness of possible parents. This provides estimated breeding values (EBVs) that predict how useful each individual will be as a parent. The effects of random and directional selection on estimates of heritability and estimated breeding values were already used for grapes (Wei et al. 2003; Smith et al. 2013). Although estimated breeding value is a good method for evaluation, genomic selection is now being used as a selection strategy, employing marker-assisted selection (MAS) based on a few identified quantitative trait loci (QTL). This approach represents an innovative shift in the assessment of plant improvement, and it is becoming possible to evaluate some plant traits based on genomic estimated breeding value (Viana et al. 2016).

One of the most important viticultural characteristics of a grapevine rootstock, for example, is its ability to form roots on dormant, lignified canes (rootstrike) (Smith et al. 2013). Rootstocks that are easy to root and graft are more economical to produce. Such rootstocks are far more widely planted than their inherent value warrants (Galet 1979). North American species of *Vitis* are the primary source of germplasm for grapevine rootstocks and vary widely in their rates of rootstrike. Breeders hybridize grape species to introgress traits to produce commercial rootstocks. A combination of parents consisting of improved and wild accessions of *Vitis* spp. are used to generate new families. The percentage of rootstrike of dormant canes can then be observed over several years to identify elite individuals. A logistic generalized linear mixed model method can be used to estimate the narrow-sense heritability (h^2) of rootstrike. The model includes estimates of the breeding values of all parents and progeny. Hence, this model can be used to estimate breeding values to identify individuals with commercially acceptable rates of rootstrike with a defined probability of transmitting this trait to progeny. This is useful for the introgression of traits into potentially new commercial rootstocks. The pattern of the normal distribution of rooting indicates that it is possible to identify individuals with good rootstrike from *Vitis* species that are generally considered to have low rootstrike. The selection of individuals with a higher breeding value will increase the efficiency of rootstock breeding (Smith et al. 2013). Vines with high estimated breeding values are retained in the breeding program, as it is predicted that these genotypes have a higher probability of producing progeny with a high rate of rootstrike (Clingeleffer and Davis 2012). One of the best methods to facilitate the evaluation of root formation in the greenhouse (nursery) is to plant cuttings in "rootrainer" containers which allow evaluation of root growth and development without causing root damage (Rahemi 2016).

Screening for K uptake can be determined by the K^+ content of dry petiole weight from the leaf opposite the primary fruit cluster from ungrafted vines at anthesis time (Rühl 1989). It is also suggested by Rühl (1990) that ungrafted vines in a greenhouse could be grown in hydroponic solution with a known level of K^+ in the media to observe the level of accumulation in the petioles. In a two years study of grafting 'Norton' on own-root, 3309 C, 101–14 Mgt, Schwarzmann, 5BB, SO4, 1103 P, 110 R, 140 Ru, 1616 C, and 44–53 M it was shown that Norton petiole contents of Ca and P were deficient on some rootstocks. Calcium was deficient in grapevines on 101–14 Mgt, 1103 P, 140 Ru, and 1616 C rootstocks. Phosphorus was inadequate in vines on 5BB and 44–53 M rootstocks and own-rooted vines in both years and in vines on 3309 C, 101–14 Mgt, 140 Ru, and 1616 C in the second year of the study. In contrast, petiole K was excessive for vines on 101–14 Mgt, and 1616 C rootstocks in the first year of the study. Although generally 'Norton' is grown on its own-roots in Missouri (U.S.), vines with 101–14 Mgt, 110 R, or 1616 C rootstocks have shown greater yield than their own-rooted counterparts (Harris 2013).

The rootstock influence on nutrient content in grape petioles was similarly evaluated by Dalbó et al. (2011) in Brazil. 'Niagara Rosada', 'Concord', and 'Isabella' were grafted on different rootstocks (VR O43-43, VR O44-4, SO4, 420A, 99 R, 1103 P, IAC 766, IAC 572, Concord, Isabella, Dogridge, Schwarzmann, 101–14 Mgt, 3309 C). The greatest rootstock effect was observed on K and Mg content and K/Mg ratio in the petioles. The *Vitis vinifera* × *V. rotundifolia* hybrid rootstocks, VR O43-43 and VR O44-4 provided the highest K/Mg values and self-rooted Isabella had the lowest K/Mg ratio. The other tested rootstocks resulted in intermediate values. There was also a significant effect on P content, but only in 'Niagara Rosada' and 'Concord'. Their results indicate that rootstock must be considered for nutritional status evaluation and fertilizer recommendation regarding K and Mg.

4.2 Scion-Rootstock Interactions and Effects

Rootstock selection can affect scion characteristics in a number of ways. These include yield (vine's capacity to balance between vegetation and fruit (hormonal influence), number of clusters per shoot, number of berries per cluster, size of berries, and condition of berries at harvest (water uptake)). Furthermore, yield, in turn, affects ripening time (due to crop load), sugar/alcohol levels, acid balance, tannin content, color, quality factors: juice concentration, complexity; canopy size (vigor from N uptake), from more or less laterals, from water uptake, from a balance of vegetation and fruit (hormones); canopy shading, in turn, affects (rot, mildew, pH/potassium, malic acid, herbaceousness); chemical composition of fruit (N uptake, K uptake, Ca uptake, other elements); chemical composition affects (rate of wine fermentation, stability, color, flavor); vegetative cycle (phenology) (bud fruitfulness affected by temperature, flower date regard to spring frost, ripening date before or after rains, ripening in warmer or cooler conditions); vine longevity (better pest and disease resistance, better winter hardiness); older vines mean (better economics if grape variety is certain, more consistent, wine character, more complex wine) (Johnson 2015). While there is no question that rootstocks influence wine flavors, there are plenty of questions as to how, when, and where this occurs (Morton, 2007). The relationships between the rootstocks and the scion are summarized in Figs. 4.1 and 4.2.

Cold tolerance (early acclimation & late de-acclimation)

Drought tolerance

Berry colour and quality

Berry size

Veraison, ripening and precocity

Yield and productivity

Diseases resistance

Vigor, dwarfing, plant architecture

Vitis vinifera

North American species

Abiotic stress tolerance (Drought, cold, …)

Special conditions (Lime, pH, bad drainage, salinity, compact soil)

Pest and Disease resistance (Phylloxera, nematodes & agrobacterium)

Fig. 4.1 Scion/rootstock relationships (Rahemi 2016)

Fig. 4.2 Interaction of scion, rootstock, and graft union in a fused grapevine. The blue zone shows the hypothetical combination of scion, rootstock, and graft union, which provides the highest productive vines. The key point is to find the superior combination (scion and rootstock) for each kind of growing area (Rahemi 2016)

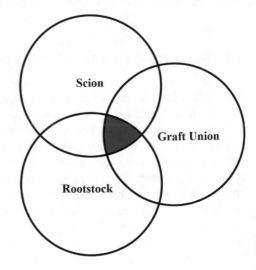

4.3   Grafting Compatibility and Affinity

Successful grafting or compatibility depends on the formation of a graft union between scion and rootstock. The inability to produce a graft union between scionand rootstocks is termed 'incompatibility' (Gargin and Altindisli 2014).

Scion rootstock relationships are extremely specific and depend on the affinity and compatibility of the grafting combination and soil and climatic adaptation (Verdugo-Vásquez et al. 2021). The compatibility and long-term affinity of the graft combination should be considered for grafting a scion onto a rootstock (van Schalkwyk 2017).

Gargin and Altindisli (2014) did not distinguish between affinity and graft compatibility in grapes, while Rives (1971) mentioned that there are clear distinctions between them due to different origins. He stated vigor has additive and non-additive components. Additive component is contributed by the rootstocks to variations in vigor, and affinity is the non-additive component of the variations (Rives 1971).

Depending on the rootstock selection and scion cultivars, incompatibilities sometimes take place (Jacob 1943; Read and Gu 2003). While some scion varieties show very good to good compatibility with rootstocks, certain varieties have shown moderate to poor compatibility (Southey 1992; Hunter et al. 2016; van Schalkwyk 2017) (Table 4.1). Rootstock 1202C shows excellent compatibility with Chardonnay scions, and it was a vigorous, high-yielding stock, while the union of rootstock 110 R with Chardonnay is conspicuous because of scion overgrowth, and the diameter of scion and rootstock in the union is different (Foott et al. 1989). Similarly, Chardonnay on 101–14 Mgt shows low compatibility, low rate of rooting in the nursery, and low survivability after planting in the field in the first season (Rahemi 2016). Not only does the rootstock and scion relationship affect graft compatibility, but different grafting methods can also influence the compatibility of the graft union and vine survival. For example, graft failures using the field grafting method were about 5%,

Table 4.1 Graft compatibility and long-term affinity of selected rootstocks in South Africa (Southey 1992; Hunter et al. 2016; van Schalkwyk 2017)

Rootstock	Affinity
99 R	Excellent
101–14 Mgt	Fair
110 R	Excellent
140 Ru	Excellent
143-B Mgt	Good
420-A Mgt	Good
775 P	Fair
1045 P	Fair
1103 P	Excellent
3306 C	Fair
3309 C	Good
Jacquez	Excellent
Ramsey	Poor
Rupestris du Lot	Fair
SO4	Good
USVIT 8–7	Good

Table 4.2 Comparison of graft compatibility with different grafting methods (Hayne 1896)

Scion/Rootstocks	Grafting method	Percent of failed by 5th year
Zinfandel/Rupestris 'St. George'	Field grafting	32
Zinfandel/3309 C	Field grafting	25
Zinfandel/101–14 Mgt	Field grafting	16
Zinfandel/Rupestris 'Martin'	Field grafting	45
Zinfandel/Vialla	Field grafting	30
Mondeuse/Rupestris 'St. George'	Bench grafting	3
Tokay/Rupestris 'St. George'	Bench grafting	4
Ferrara/Rupestris 'St. George'	Bench grafting	7

while with bench grafting, it was about 30% (Hayne 1896) (Table 4.2). The rootstock parentage species can also impact grafting compatibility. For example, while *V. riparia* as a rootstock showed good grafting ability, *V. berlandieri* has a poor grafting yield. Furthermore, the graft yield is very variable across different genotypes of *V. riparia* (Rahemi 2016).

References

Christensen LP (2003) Rootstock selection. In: Bettiga LJ (ed) Wine grape varieties in California. University of California, Oakland, CA, USA, pp 12–15

Clingeleffer P, Davis H (2012) Rootstock breeding and development for Australian dried grapes. CSIRO plant industry, Horticulture Australia Ltd., Canberra

Dalbó MA, Schuck E, Basso C (2011) Influence of rootstock on nutrient content in grape petioles. Rev Bras Frutic, Jaboticabal–SP 33(3):941–947

Ferreira JHS (1985) Effect of rootstock on the incidence of dying arm of Chenin blanc vines. S Afr J Enol Vitic 6(1):23–24

Foott JH, Ough CS, Wolpert JA (July–Aug 1989) Rootstock effects on wine grapes. Calif Agric 27–29

Galet P (1979) A practical ampelography, Translated and adapted by Lucie T. Morton. Ithaca, NY: Comstock Cornell University Press

Gargin S, Altindisli A (2014) A research on the affinity coefficients of red globe grape variety with 140 R, 41 B rootstocks. In: BIO web of conferences, vol 3. EDP sciences, p 01004

Harris JL (2013) Effect of rootstock on vegetative growth, yield, and fruit composition of Norton grapevines. MSc Thesis, University of Missouri, p 95

Hayne P (1896) Resistant vines, their selection, adaptation, and grafting, appendix to viticultural report, University of California, College of Agriculture, Agricultural experiment station

Hunter JJ, Archer E, Van Schalkwyk D, Strever AE, Volschenk CG (2016) Grapevine roots: interaction with natural factors and agronomic practices. Acta Hortic 1136:63–80

Jacob HE (1943) Examples of incompatibility between grape varieties and rootstocks. Proc Ame Soc Hort Sci 41:201–203

Johnson A (2015) Phylloxera and rootstocks, a viticulture course document, Napa Valley college, reading materials, pp 25–73. http://www.napavalley.edu/people/ajohnson/Pages/ajwelcome.aspx

Morton LT (2007) Rootstock review, how to set priorities when choosing rootstocks, wines and vines

Pool B, Lerch S, Howard G, Johnson T, Weimann D (2005) Rootstocks for planting or replanting New York vineyards, Finger lakes vineyard notes. Cornell University, pp 1–14

Rahemi A (2016) Breeding of grapevine rootstocks for adaptability to the environmental stresses, third national conference of grape and raisin conducted by research institute of grape and raisin (RIGR), 27–29 September 2016. Malayer University, Malayer, Iran

Read PE, Gu S (2003) A century of American viticulture. HortScience 38(5):943–951

Reynolds AG, Wardle DA (2001) Rootstocks impact vine performance and fruit composition of grapes in British Columbia. HortTechnol 11:419–427

Rives M (1971) Statistical analysis of rootstock experiments as providing a definition of the terms vigour and affinity in grapes. Vitis 9:280–290

Rühl EH (1989) Uptake and distribution of potassium by grapevine rootstocks and its implication for grape juice pH of scion varieties. Aus J Exp Agric Anim Hus 29:707–712

Rühl EH (1990) Better rootstocks for wine grape production. Aust Grape Grower Wine Maker 316:113–115

van Schalkwyk D (2017) Fruit production: get the right rootstock for planting grapes, 14 Dec 2017. Retrieved: 9 Sep 2019. https://www.africanfarming.com/fruit-production-get-right-rootstock-planting-grapes

Smith BP, Wheal MS, Jones TH, Morales NB, Clingeleffer PR (2013) Heritability of adventitious rooting of grapevine dormant canes. Tree Genet Genomes 9(2):467–474

Southey JM (1992) Grapevine rootstock performance under diverse conditions in South Africa. Paper presented at: Rootstock seminar: a worldwide perspective, Reno, Nevada, USA

Verdugo-Vásquez N, Gutiérrez-Gamboa G, Díaz-Gálvez I, Ibacache A, Zurita-Silva A (2021) Modifications induced by rootstocks on yield, vigor and nutritional status on *Vitis vinifera* Cv Syrah under hyper-arid conditions in Northern Chile. Agronomy 11(5):979. https://doi.org/10.3390/agronomy11050979

Viana AP, Vilela de Resende MD, Riaz S, Walker MA (2016) Genome selection in fruit breeding: application to table grapes. Scientia Agricola 73(2):142–149

Wei X, Sykes SR, Clingeleffer PR (2003) Effects of selection on early stage genetic evaluation for berry weight in breeding table grapes. Plant Breed 122(1):77–82

Chapter 5
Biotic Stresses and Rootstock Selection

5.1 Phylloxera

Grape phylloxera (*Daktulospharia vitifoliae* Fitch) is a gall-forming (cecidogenic) aphid-like belonging to the family Phyllocridae. It is native to the Eastern United States of America. The host range of phylloxera is limited to members of the genus *Vitis,* which comprises species native to America, Asia, and Europe. It lives in pouch-like galls (gallicola) on the leaves of some American species and overwinters on the roots where galls (radicicola) may form as simple swellings. Feeding results in the formation of two different types of root swelling: nodosity and tuberosity (Kerbs 1995). Tuberosities (feeding sites on larger, structural roots with periderm) are always found with nodosities, but nodosities can occur alone and are found on phylloxera tolerant species.

The roots of the American species show varying degrees of tolerance to phylloxera, and many species can survive infestation without being adversely affected. The response to phylloxera of *V. vinifera,* the vine species involved in most commercial grape production, is quite different from that of American *Vitis* species. On *V. vinifera*, phylloxera is generally unable to form leaf galls, but the roots have no effective defense mechanism and become heavily infested, with the formation of many well-developed root galls. The formation of galls on primary roots (tuberosity) of susceptible vines leads to decay of the root system and eventual death of the vine. Once it became apparent that there was no available method for eradicating phylloxera, other means of maintaining commercial production of grapes in infested areas were sought. Laliman (1869) firstly suggested grafting *V. vinifera* onto the roots of various tolerant or resistant American *Vitis* species. Grafting proved successful, and many American *Vitis* species and their hybrids, showing varying degrees of tolerance or resistance to phylloxera, are used as rootstocks. Most rootstocks currently in use are only phylloxera tolerant rather than resistant. Börner (a hybrid of *V. cinerea* and *V. riparia*) was introduced as one of the resistant rootstocks. This rootstock reacts with a hypersensitive reaction (HR), leading to local necrosis on leaves and roots, thus preventing the feeding and breeding of the insect (Börner 1943; Niklowitz 1955;

Anders 1958; Schmid et al. 2015). In a cross-breeding program at the Geisenheim Institute (Germany), *V. cinerea*, is used for the crosses, and each of the individuals is inoculated with phylloxera leaf galls. Then in the greenhouse, they evaluated their leaf resistance by measuring the production of a hypersensitive response. Approximately 1% of the individuals show leaf resistance (Dry 2005). The identification of admixed samples among native grape phylloxera populations indicates that shared environments facilitate sexual reproduction between different host associated populations to create new genotypes of phylloxera (Lund et al. 2017). Phylloxera from different sources varied in the number of generations per season as a function of rootstock (Corrie et al. 1998). Phylloxera damage and vine growth assessment is evaluated in terms of levels: 0 = dead; 1 = extremely poor vine, almost dead; 3 = poor growth; 5 = weak to average growth; 7 = normal, healthy growth; 9 = very good growth (Porten et al. 2000). Assessment of phylloxera populations are designated into one of several categories. Roots are visually assessed and assigned to 1 to 9 as detailed below.

1 = no infestation: no nodosities and no phylloxera;
3 = low infestation: few new nodosities, very low phylloxera population, only a few eggs;
5 = medium infestation: several new nodosities, many old ones, some phylloxera, some eggs;
7 = heavy infestation: many new and old nodosities, many phylloxera, often several on one nodosity (feeding site on secondary root swellings that occur near the root tips), many eggs; or
9 = extreme infestation: lots of new and old nodosities, a large number of phylloxera and eggs, usually several adults on one nodosity (Porten et al. 2000).

Rootstock evaluation is done based on the reaction of hybrid leaves and roots to phylloxera attack and ranking on a scale from 1 to 9 as well and is detailed below.

1 = no galls: no nodosities;
3 = galling without egg production: root gall like structures, no phylloxera;
5 = small galls: small nodosities;
7 = gall production: normal nodosities;
9 = extreme gall production: a large number of nodosities, adult phylloxera and eggs (Porten et al. 2000).

Other researchers, such as Viala and Ravaz (1903), have used more detailed ratings covering a 0–20 scale. They claimed that species or vines with ratings of 16–20 are adequately resistant to phylloxera for use in all soil types, that ratings of 14 and 15 are sufficient for sandy and damp soils, and that ratings below 13 have insufficient resistance and should not be used commercially. This classification is including: 20 = Immunity, a term that we consider inappropriate considering the rise of diverse of phylloxera biotypes: 19 = *V. rotundifolia*; 18 = *V. cordifolia*, *V. rupestris*, *V. arizonica*, *V. riparia*, *V. rubra*, *V. riparia* × *V. rupestris*; 17 = *V. berlandieri*, *V. monticola*, *V. berlandieri* × *V. riparia*, *V. berlandieri* × *V. rupestris*; 16 = *V. rupestris* 'St. George' ('Rupestris du Lot'), *V. aestivalis*; 15 = *V. mustangensis* (*V. candicans*), *V. cinerea*;

14 = *V. lincecumi, V. champinii, V. longii* (*V. solonis*); 13 = *V. doaniana*; 12 = *V. glabrous* × *V.* × *champinii*, 'Jacquez' ('Lenoir'); 10 = 'Elvira'; 6 = 'Othello'; 5 = *V. labrusca*; 4 = 'Catawb'a; 3 = 'Delaware', 'Concord', *V. coignetiae*; 2 = 'Rulander' ('Pinot noir'), *V. amurensis*; 1 = 'U'ngi blanc'; 0 = 'Cabernet Sauvignon' (Walker 1991).

Boubals (1966) also classified rootstocks for phylloxera tolerance. The Boubals scale of three classes was based on root studies from phylloxera-infested vines grown in pots. The formation of tuberosities and subsequent decay or necrosis was the basis of assessment. Boubals (1966) recommended that only rootstocks with phylloxera resistance ratings of 0 or 1 be used commercially (Class 0 = no indication of phylloxera activity on the roots of the plant; eg., *V. rotundifolia*, some *V. riparia, V. rupestris, V. berlandieri,* and *V. cordifolia*, 420A, 779P, 1616 C; Class 1 = from a series of inoculated plants, one or two tuberosities appear, and signs of necrosis on the roots; e.g., *V. riparia, V. rupestris, V. berlandieri, V. cordifolia*, 3306 C, 3309 C, 101–14 Mgt, 33EM, 5BB, 5C, SO4, 8B, 157–11, 161–49, 775P, 1103 P, 140 Ru, 99 R, 110 R, 41B, AXR#2, 44–53 M, Salt Creek; Class 2 = several tuberosities in chains, or a few tuberosities but with definite necrosis of the root; eg., *V. labrusca, V. monticola*, some selections of *V. riparia, V. rupestris, V. berlandieri*, 333EM, AXR#1, AXR#9, Catawba, Dog Ridge; Class 3 = large portion of the roots destroyed and decomposed as a result of phylloxera feeding; e.g., *V. arizonica, V. girdiana, V. vinifera*, some *V. labrusca*, 1613C (Walker 1991).

Pàstena (1976) classified grape rootstocks based on a 0 (susceptible)—10 (resistant) scale of root damage in potted plants Variable susceptibility to phylloxera was seen in *V. vinifera* cultivars. The ranking was: 10 = *V. rotundifolia*; 9 = *V. riparia, V. berlandieri*, and *V. cordifolia*; 8 = *V. rupestris*, 779 P, 420A, 41B; 7 = St. George, 1103 P, 140 Ru, 110 R, 775P, 41B (in Palermo), 1447 P; 6 = 3306 C, 3309 C, 225 Ru, SO4, 5BB, 161–49 C, 157–11, 1045 P, 44–53 M; 5 = *V. solonis*, 'Jacquez'; 4 = AXR#1, several hybrids direct producers; 2–3 = several *V. vinifera* cultivars: 'Tannat', 'Nocera', 'Olivetta', 'Carignane', 'Sangiovese', 'Barbera', 'Trebbiano dorato', 'Sauvignon blanc', 'Grecanico'; 1 = hybrid direct producers (many Seyve-Villard numbers, some Seibel); 0 = *V. vinifera* and more hybrid direct producers (Walker 1991).

Phylloxera resistant attributes of the most popular rootstocks was evaluated in South Africa (Southey 1992; Hunter et al. 2016; van Schalkwyk 2017). Rootstocks were classified as resistant, moderately resistant, moderately susceptible, and susceptible. The rootstock 99 R was shown to be exceptional with very good resistance to phylloxera (Table 5.1).

Different biotypes (a homogenous population showing distinct biological characteristics) and strain (a homogenous population showing distinct genetic characteristics) of phylloxera populations have been identified (Kellow 2000). DNA typing has been used to distinguish phylloxera biotypes in Australia, some of which were geographically distinct. Additional work described up to 83 biotypes. Given that no evidence of sexual reproduction has been found in Australia, it is speculated that these biotypes were either brought into Australia or that mutations have occurred (Corrie et al. 1998, 2001, 2002). Similar results have been documented in Europe,

Table 5.1 Phylloxera resistant attributes of the most popular rootstocks used commercially in South Africa (Southey 1992; Hunter et al. 2016; van Schalkwyk 2017)

Rootstock	Phylloxera resistance
99 R	Resistant
101–14 Mgt	Moderately resistant
110 R	Resistant
140 Ru	Moderately resistant
143-B Mgt	Moderately resistant
420-A Mgt	Moderately susceptible
775 P	Resistant
1045 P	Resistant
1103 P	Resistant
3306 C	Moderately resistant
3309 C	Resistant
Jacquez	Susceptible
Ramsey	Resistant
Rupestris du Lot	Resistant
SO4	Moderately resistant
USVIT 8–7	Moderately resistant

where one report identified 103 biotypes, that sexual reproduction was rare (possibly before the introduction to Europe), and migration rates between populations were low (Vorwerk and Forneck 2006). Some biotypes in Australia live exclusively on the leaves (e.g., G52, G54), some exclusively on the roots (e.g., G1, G4, G39, G51), and others live on both roots and leaves (e.g., G2, G3, G35, G53, G56) (Corrie and Hoffmann 2004). Biotypes A and B also have been reported in Australia. Biotype B as being biologically distinct from a previously characterized isolate of phylloxera, biotype A. Biotype B has increased fecundity and shorter generation time on the rootstock. Most importantly, it can colonize secondary roots and thus overcome the tolerance of this rootstock in the field, which biotype A is unable to do. Distinct biotypes have now been characterized for a range of additional phylloxera strains, including strains 1, 2, 3, and 4 from California, a German strain, Hungarian strains Hung 1, 2, and 3. In Australia, four genetically distinct populations have been identified. Two main biotypes have been described: one like biotype B, which includes strain VWL-1, and a second one distinct from either biotype A or B. This biotype, which includes strain SRU-1, is less damaging to *V. vinifera* than VWL-1 and has higher survival and developmental rates on the rootstocks than on *V. vinifera* (Kellow 2000).

Table 5.2 Provisional resistance ratings of rootstocks to phylloxera biotypes (Powell 2009)

Rootstock	Phylloxera biotype					
	G1	G4	G7	G19	G20	G30
V. vinifera	S/S	S/S	S/S	S/S	S/S	S/S
Ramsey	T/T	T/T	T/T	T/T	T/T	T/T
Schwarzmann	R/R	R/R	T/T	T/T	T/T	T/T
Börner	R/R	R/R	T/R	R/R	R/R	T/R
110 R	T/T	T/T	T/R	R/R	R/R	T/R
1103 P	T/T	T/T	R/nd	T/T	R/nd	R/T
140 Ru	T/T	T/T	R/R	R/R	R/nd	R/R
5BB Kober	R/R	R/R	T/T	T/T	T/T	T/T

S = susceptible, T = tolerant, R = resistant, nd = not determined (first letter = excised root assay/second letter = potted vine assay)

In a special method, phylloxera and grapevines were co-cultivated using both potted and micropropagated grapevines. The development of nodosities on primary roots was studied by microscopy and histochemistry, and nodosities were analyzed for biochemical changes and changes in gene expression. The results showed the nodosities on *V. vinifera* potentially function as nutrient reservoirs, and defense responses to phylloxera attacks were not detected (Kellow et al. 2004).

Powell (2009) compared provisional resistance ratings of rootstocks (*V. vinifera*, Ramsey, Schwarzmann, Börner, 110 R, 1103 P, 140 Ru, and 5BB) to phylloxera biotypes (G1, G4, G7, G19, G20, and G30) and classified them as being susceptible, tolerant or resistant by two methods (excised root assay, and potted vine assay), Börner was one of the rootstocks which showed resistance to all biotypes (Table 5.2).

Progressive Viticulture group (2011) recommends replanting a phylloxera-infested vineyard with a rootstock selected from a different group than that used currently in the vineyard. The specified groups were: group A (*V. berlandieri* × *V. riparia* including SO4, 5BB, 5C, 125AA, and 420A), Group B (*V. berlandieri* × *V. rupestris* including 99 R, 110 R, 140 Ru, and 1103 P), Group C (*V. riparia* × *V. rupestris* including 3309 C, 101–14 Mgt, and Schwarzmann), Group D (Selection or other hybrids including St. George, RGM, 1616 C, 44–53 M, Freedom, and Harmony), and Group E (*V.* × *champinii* including Ramsey and Dogridge) (Table 5.3).

Pavloušek (2012), studying 59 grape rootstock hybrids originating from nine different populations, found that the highest number of resistant hybrids was within a Binova × Börner family. Börner was found to be the best gene source of phylloxera resistance (Pavloušek 2012).

Table 5.3 Phylloxera replant rootstock groups (Progressive Viticulture 2011)

Replant group	Group A	Group B	Group C	Group D	Group E
Parentage	*V. berlandieri* × *V. riparia*	*V. berlandieri* × *V. rupestris*	*V. riparia* × *V. rupestris*	Selection or other hybrids	*V.* × *champinii*
	SO4	99 R	3309 C	St. George	Ramsey
	5BB	110 R	101–14 Mgt	RGM	Dogridge
	5C	140 Ru	Schwarzmann	1616 C	
	125AA	1103 P		44–53 M	
	420A			Freedom	
				Harmony	

*Before replanting a phylloxera-infested vineyard, select a rootstock from a group that is different from the group to which the rootstock currently in your vineyard belongs (Copyright © 2011 Progressive Viticulture, http://www.progressivevit.com/selecting-a-rootstock-for-a-wine-grape-vineyard/)

5.2 Nematodes

Nematode infection not only can transmit certain viruses but also can affect water stress, nutrient deficiency, stunting of shoots, and decreased yield. The most important nematode arguably is *Xiphinema index*, one of the two Dagger nematodes, as it vectors grapevine fanleaf virus. *Xiphinema americanum* is the other dagger nematode and although it is widespread in the states and can vector tobacco ringspot virus, it is typically less of a concern in viticulture. Aside for *X. index*, arguably the next most impactful nematode to commercial grapevines are the root-knot nematodes (RKN) (*Meloidogyne incognita, M. arenaria, M. hapla, M. javanica,* and *M. chitwoodi*). Ring nematode (*Mesocriconema xenoplax*), root lesion nematode (*Partylenchus vulnus*), citrus nematode (*Tylenchulus semipenetrans*), needle nematode (*Longidorus africanus*), and sheath nematode (*Hemicriconemoides* spp.) also are moderately important on grapes (UC IPM 2021).

The United States Department of Agriculture (USDA) has developed methods for the rapid screening of germinated seedlings inoculated with nematodes. Around 1% of seedlings show resistance to aggressive root-knot nematodes (Whiting 2012).

Stamp (2011) reported that among 19 rootstocks (Freedom, 1613C, Harmony, 5C, SO4, Schwarzmann, VR O39-16, 3309 C, 101–14 Mgt, 5BB, 110 R, RS-2, RS-3, RS-9, GRN-1, GRN-2, GRN-3, GRN-4, and GRN-5), only GRN-1 was resistant to all important nematodes (*M. incognita, X. index, X. americana, Mesocriconema xenoplax,* and *Partylenchus vulnus*. Rootstocks were divided into different classes, including resistant, slightly susceptible, high susceptible, moderately resistant, and susceptible (Table 5.4).

Progressive Viticulture (2011) categorized rootstocks into five groups and recommended the use of rootstocks from a different group when replanting a root-knot nematode-infested vineyard (Table 5.5).

Table 5.4 Relative resistance/susceptibility of rootstock to nematodes (Adapted from Stamp 2011)

Rootstock	Root Knot *M. incognita*	Dagger *X. index*	Dagger *X. americanum*	Ring *M. xenoplax*	Lesion *P. vulnus*
Freedom	R	R	MR	HS	SS
1613 C	R	MR	HS	HS	SS
Harmony	MR-R	MR	HS	MS	SS
5C	SS-MR	MR	HS	HS	S
SO4	SS-MR	MR	S	HS	S
Schwarzmann	S-MR	MR	SS	MS	SS
O39-16	S	R	MR	NA	S
3309 C	HS-S	SS	S	HS	SS
101–14 Mgt	MR	HS	NA	HS	S
5BB	S	HS	NA	HS	S
110 R	R	HS	NA	HS	S
RS-2	SS	HS	NA	HS	R
RS-3	R	MR	MR	MS	na
RS-9	R	MR	MR	HS	na
GRN-1	R	R	R	R	R
GRN-2	R	R	R	S	R
GRN-3	R	R	R	S	R
GRN-4	MR	R	R	MS	R
GRN-5*	R	R	R	MS	R

* Does support low populations of phylloxera on roots, but this has not been shown to damage vines (R = resistant, SS = slightly susceptible, HS = high susceptible, MR = moderately resistant, S = susceptible, NA = not available)

Table 5.5 Root-knot nematode replants rootstock groups (Progressive Viticulture 2011)

Replant Group	Group E	Group F	Group G	Group H	Others
Parentage	*V. champinii*	1613 × *V. champinii* (DR)	*V. champinii* (R) × Schwarzmann	*V. champinii* with other species	Other complex hybrids
	Ramsey	Freedom	RS-3	Kingfisher	Matador
	Dogridge	Harmony	RS-9	GRN-2	Minotaur
				GRN-3	GRN-1
				GRN-4	

Clingelffer and Smith (2011) examined 69 rootstocks for their resistance to *Meloidogyne javanica* in Australia. They categorized them as resistant (39 rootstocks), tolerant (7 rootstocks), and susceptible (23 rootstocks). Ferris et al. (2012), in a review report on 27 grape rootstocks resistance to 10 nematodes (*M. incognita* Race 3, *M. javanica, Meloidogyne* pathotypes Harmony A&C, *M. chitwoodi, X. index, M. xenoplax, P. vulnus, T. semipenetrans, X. americanum,* and *P. hamatus*), showed that no rootstock expressed resistance to all nematodes. Resistance assessed relative to nematode reproduction on 'Colombard' (or another susceptible cultivar) showed 10% were resistant, 10–30% moderately resistant, 30–50% moderately susceptible, and >50% were susceptible (Table 5.6).

A compilation based on some studies which presented by (Anwar et al. 1999, 2002; Chitambar and Raski 1984; Gu and Ramming 2005a, b; McKenry et al. 2001a, b). The case of *P. vulnus,* has included some observations on host status based on tissue culture plantlets. Resistance assessed relative to nematode reproduction on cv Colombard R = resistant, MR = moderately resistant, MS = moderately susceptible, S = susceptible.

Dry (2008) reported the resistance of 12 rootstocks to four nematodes (*Meloidogyne* spp., *T. semipenetrans, Partylenchus* spp., and *Xiphinema* spp.) with Ramsey and Schwarzmann showing the most resistance (Table 5.7).

Ramsey showed very good nematode resistance attributes compared with the other most popular rootstocks in South Africa (Southey 1992; Hunter et al. 2016; van Schalkwyk 2017) (Table 5.8).

Walker et al. (2018) evaluated resistance to nematodes on five newly released rootstocks (GRN-1, GRN-2, GRN-3, GRN-4, and GRN-5). They used different nematodes in a combined inoculation and screened for resistance to lesion nematode, citrus nematode, and ring nematode. GRN-1 and GRN-5 showed the most resistance to different nematodes. GRN-2, GRN-3, and GRN-4 were not resistant to ring nematode. GRN-2 was also moderately susceptible to citrus nematode (Table 5.9).

5.3 Viruses

Grapevines can be affected by many viruses and viral diseases, and can induce a variety of different problems in grapevines, including the degeneration/decline disease complex, Leafroll disease complex, Rugose wood disease complex, Fleck disease complex (Martelli et al. 2007; Oliver and Fuchs 2011). A list of grapevine viruses, their vector, and the disease they cause have been presented by Martelli (2018), and updated by Fuchs (2020) (Table 5.10).

However, despite a long history of cultivation and breeding efforts, little usable resistance to viral diseases has been identified in *Vitis* species. As a result, the management of viral diseases has largely relied upon prevention/exclusion or on the use of economically and/or environmentally consequential methods to manage vector populations. Resistance to dagger nematode vectors has been identified and successfully transferred to rootstock genotypes that are commercially available, but these resistant

Table 5.6 Host status of grape rootstocks to nematodes (Ferris et al. 2012)

Genotype	M. incognita, Race3	M. javanica	Meloidogyne pathotypes Harmony A&C	M. chitwoodi	X. index	M. xenoplax	Pra. vulnus	T. semipenetrans	X. americanum	Para. hamatus
101–14 Mgt			R		S	S	MR			S
1103 P			S		S	S	MS			S
110 R			MR		S	S	S			S
140 Ru			MR		S	S	S			MS
1613 C	R	R	S	R	MR	S	MS	S	S	
1616 C			MR		S	S	MS			S
3309 C	S	S	S		MS	S	S	S	S	S
420A			R		S	S	MS			S
44–53 M			S		S	MR	MS			S
A × R1			S		S	S	S			S
Börner			R		R	S	MS			
Dog ridge	R	R	S		S	S		MR	MR	MS
Freedom	R	R	S	S?	R	MS	MS	S	MS	MR
Harmony	R	R	S	S	MS	S	S	S	S	S
K51-32	MR				MS	S	R			S
5BB			R		S	S	MS			MR
Ramsey0	R	R	S	S?	MR	S	MS	MSS	S	S
RGM			R		R	S	MR			S
RS-3	R	R	MR	MR	S	S	MR			S
RS-9	R	R	R	R	S	S	MS			S
Schwarzmann	S	MR	S		MR	MS	S	S	MS	S

(continued)

Table 5.6 (continued)

Genotype	M. incognita, Race3	M. javanica	Meloidogyne pathotypes Harmony A&C	M. chitwoodi	X. index	M. xenoplax	Pra. vulnus	T. semipenetrans	X. americacanum	Para. hamattus
St. George	S		S		S	S	MS	S	S	MS
5C	MS	MR	S		MR	MS	S	R		MS
USDA10-17A	R	R	R	R	R	MS	R	R		
USDA10-23B	R	R	R	R	R	MR	R	R		
USDA6-19B	R	R	MS	R	MR	MR	R	R	R	
VR O39-16	S	S	S		R	R	MR	S	MR	MR

Table 5.7 Nematode resistance (Nicol et al. 1999; Whiting 2003; Dry 2008)

Rootstock	Root Knot *M.* spp.	Citrus *T. semipenetrans*	Root lesion *P.* spp.	Dagger *X.* spp.
Ramsey	High	High	High	Low
Schwarzmann	High	High	Low	High
101–14 Mgt	High	Low	Low	Low-Moderate
5C Teleki	Moderate	Low	Moderate	Low
5BB Kober	High	Low	Low	Low
SO4	Moderate-High	Low	Moderate	Low
1103 P	Low	Moderate	Moderate	Low
110 R	Moderate	Moderate	Low	Low
99 R	Moderate-High	Low	Low	Low
140 Ru	High	–	Low	Low
3309 C	Low	Low	Moderate	Moderate
420 A	Moderate	Low	Low	Low

Table 5.8 Nematodes resistant attributes of the most popular rootstocks in South Africa (Southey 1992; Hunter et al. 2016; van Schalkwyk 2017)

Rootstock	Nematodes
99 R	Moderately resistant
101–14 Mgt	Moderately resistant
110 R	Moderately resistant
140 Ru	Moderately susceptible
143-B Mgt	Moderately resistant
420-A Mgt	Moderately susceptible
775 P	Moderately resistant
1045 P	Moderately resistant
1103 P	Moderately resistant
3306 C	Moderately susceptible
3309 C	Moderately resistant
Jacquez	Susceptible
Ramsey	Very resistant
Rupestris du Lot	Resistant
SO4	Moderately resistant
USVIT 8–7	Moderate

materials do not prevent virus translocation into the scion, although a significant delay in infection can allow for suitable production. The resistance toward both viruses and their vectors, or lack thereof, is important in disease management (Oliver and Fuchs 2011).

Table 5.9 Nematode resistance of GRN rootstock (Walker et al. 2018)

Selection	(1) Root-knot R3	(2) HarmA	(3) HarmC	(4) X. index	(2–4) Combined	High Temp	Lesion	Citrus	Ring
GRN-1	R	R	R	R	R	R	R	R	R
GRN-2	R	R	R	R	R	R	R	MS	S
GRN-3	R	R	R	R	R	R	R	R	S
GRN-4	R	R	R	R	R	R	R	R	MR
GRN-5	R	R	R	R	R	R	R	R	R

1. Root-knot R3 is a standard strain of *Meloidogyne incognita* that feeds on *V. vinifera* and is resisted by Harmony, Freedom, and many other rootstocks. 2. HarmA is a root-knot strain (*M. arenaria*) that feeds on Harmony and Freedom. 3. HarmC is a root-knot strain (*M. incognita*) capable of feeding on Harmony and Freedom. 4. Dagger nematode (*Xiphinema index*). Combined testing used HarmA, HarmC, and X. *index* in a combined inoculation. These plants were also tested at high soil temperatures (30 °C), which "break" the root-knot nematode resistance in Harmony and Freedom with the combined inoculum (2 through 4). They were then screened for lesion nematode, citrus nematode, and ring nematode (*Mesocriconema xenoplax*). R = resistant, S = susceptible, MS = moderately susceptible; MR = moderately resistant

Table 5.10 Grapevine viruses, their vector, and the disease they cause (Martelli 2018; updated by Fuchs 2020)

Family[a]	Genus[a]	Species[a]	Acronym	Vector[b]	Disease[c]
Alphaflexiviridae	*Potexvirus*	*Potato virus X*	PVX	None	Unknown
Betaflexiviridae	*Fivirus*	*Grapevine Kizil Sapak virus*	GKSV	Unknown	Unknown
	Fovea virus	*Grapevine rupestris stem pitting-associated virus*	GRSPaV	Unknown	Rugose wood
		Grapevine virus T	GVT		
	Trichovirus	*Grapevine berry inner necrosis virus*	GINV	Eriophyid mite	Berry inner necrosis
		Grapevine Pinot gris virus	GPGV		Leaf mottling/deformation
	Vitivirus	*Grapevine virus A*	GVA	Mealybugs, soft scales	Rugose wood
		Grapevine virus B	GVB		
		Grapevine virus D	GVD		
		Grapevine virus E	GVE		
		Grapevine virus F	GVF		
		Grapevine virus G	GVG		
		Grapevine virus H	GVH		
		Grapevine virus I	GVI		
		Grapevine virus J	GVJ		
		Grapevine virus K	GVK		
		Grapevine virus L	GVL		
		Grapevine virus M	GVM		
Bromoviridae	*Alfamovirus*	*Alfalfa mosaic virus*	AMV	Aphids	Yellow mosaic

(continued)

Table 5.10 (continued)

Family[a]	Genus[a]	Species[a]	Acronym	Vector[b]	Disease[c]
	Anulavirus	Related to Amazon lily mild mottle virus	ALiMMV	None	Unknown
	Cucumovirus	Cucumber mosaic virus	CMV	Aphids	Unknown
	Ilarvirus	Grapevine angular mosaic virus	GaMoV	None	Angular mosaic
		Grapevine line pattern virus	GLPV		Line pattern
		Grapevine virus S	GVS		Unknown
Bunyaviridae	Tospovirus	Tomato spotted wilt virus	TSWV	Thrips	Unknown
Caulimoviridae	Badnavirus	Grapevine vein clearing virus	GYCV	Aphids	Vein clearing Unknown
		Grapevine badnavirus 1	GBV1		Roditis discoloration
		Grapevine Roditis leaf discoloration-associated virus	GRLDaV		
Closteroviridae	Closterovirus	Grapevine leafroll-associated virus 2	GLRaV2	Unknown	Leafroll/Incompatibility
	Ampelovirus	Grapevine leafroll-associated virus 1	GLRaVI	Mealybugs, soft scales	Leafroll
		Grapevine leafroll-associated virus 3	GLRaV3		
		Grapevine leafroll-associated virus 4	GLRaV4		
		Grapevine leafroll-associated virus 13	GLRaV13	Unknown	Leafroll
	Velarivirus	Grapevine leafroll-associated virus 7	GLRaV7	Unknown	Unknown

(continued)

Table 5.10 (continued)

Family[a]	Genus[a]	Species[a]	Acronym	Vector[b]	Disease[c]
Endornaviridae	Endornavirus	Grapevine endophyte endornavirus	GEEV	None	Unknown
Geminiviridae	Begomovirus	Grapevine begomovirus A	GBVA	Whiteflies	Unknown
	Grablovirus	Grapevine red blotch virus	GRBV	Treehopper	Red blotch
		Wild Vitis latent virus 1	WVV1	Unknown	Unknown
	Unassigned	Grapevine geminivirus A	GGVA	Unknown	Unknown
		Temperate fruit-decay-associated virus	TFDaV		
Luteoviridae	Enamovirus	Grapevine enamovirus 1	GEV1	Aphids	Unknown
Partitiviridae	Deltapartitivirus	Grapevine cryptic virus	GCV	None	Unknown
Phenuiviridae	Rubodvirus	Grapevine Garan dmak virus	GGDV	Unknown	Unknown
		Grapevine Muscat rose virus	GMRV		Unknown
Potyviridae	Potyvirus	Bean common mosaic virus	BCMV	Aphids	Unknown
		Potato virus Y	PVY		
Reoviridae	Unassigned	Grapevine Cabernet Sauvignon reovirus	GCSV	Plant/Leafhoppers	Unknown
Secoviridae	Cheravirus	Apple latent spherical virus	ALSV[d]	Unknown	Unknown
	Fabavirus	Broad bean wilt virus	BBWV	Aphids	Unknown
		Grapevine fabavirus	GFV		
	Nepovirus	Artichoke Italian latent virus	ACLV	Unknown	Degeneration
		Arabis mosaic virus	ArMV	Dagger nematode	
		Blueberry leaf mottle virus	BBLMV	Unknown	

(continued)

Table 5.10 (continued)

Family[a]	Genus[a]	Species[a]	Acronym	Vector[b]	Disease[c]
		Cherry leafroll virus	CLRV		
		Grapevine Anatolian ringspot virus	GARSV		
		Grapevine Bulgarian latent virus	GBLV		
		Grapevine deformation virus	GDeV		
		Grapevine chrome mosaic virus	GCMV		
		Grapevine fanleaf virus	GFLV	Dagger nematode	
		Grapevine Tunisian ringspot virus	GTRV	Unknown	
		Peach rosette mosaic virus	PRSM	Dagger nematode	
		Raspberry ringspot virus	RpRSV		
		Tobacco ringspot virus	TRSV		
		Tomato ringspot virus	ToRSV		
		Tomato black ring virus	TBRV		
	Unassigned	Strawberry latent ringspot virus	SLRSV	Dagger nematode	
Tombusviridae	Carmovirus	Carnation mottle virus	CarMV	None	Unknown
	Necrovirus	Tobacco necrosis virus D	TNV-D	None	Unknown
	Tombusvirus	Grapevine Algerian latent virus	GALV	None	Unknown
		Petunia asteroid mosaic virus	PAMV		
Tymoviridae	Marafivirus	Blackberry virus S	BlVS	Leafhoppers	Unknown
		Grapevine asteroid mosaic-associated virus	GAMaV		Asteroid mosaic

(continued)

Table 5.10 (continued)

Family[a]	Genus[a]	Species[a]	Acronym	Vector[b]	Disease[c]
		Grapevine asteroid mosaic-associated virus	GAMaV		Fein feathering
		Grapevine Syrah virus 1	GSyV1		Unknown
Virgaviridae	*Tobamovirus*	*Grapevine virga-like virus*	GYLV	None	Unknown
		Tobacco mosaic virus	TMV		
		Tomato mosaic virus	ToMV		
Unassigned	*Idaeovirus*	*Raspberry bushy dwarf virus*	RBDV	None	Yellow line pattern
	Sobemovirus	*Sowbane mosaic virus*	SoMV	None	Unknown
	Virtovirus	*Grapevine virus satellite*	GV-Sat	Beetles	Unknown
	Unassigned	*Grapevine Ajinashika virus*	GAgV	Unknown	Unknown
		Grapevine labile rod-shaped virus	GLRSV	Unknown	Unknown
		Grapevine stunt virus	GSV	Leafhopper	Unknown

[a] Some of the taxonomic affiliations and virus names are tentative, as ratifications by the International Committee on Taxonomy of Viruses are pending

[b] Some of the vectors are deduced by analogy with vectors of other viruses of the same genus rather than from conclusive transmission assays

[c] Most virus disease symptoms cannot be attributed to a single virus species, as symptomatic vines are predominantly mixed infected in the vineyard, unless Koch's postulates have been fulfilled

[d] Apple latent spherical virus infects experimentally grapevine seedlings and tissue-cultured grapevines as a virus vector (Maeda et al. 2020). This virus was not identified in naturally infected vines

Studies of rootstock varieties have shown different tolerances to virus and virus-like diseases: fan-leaf, leafroll, and stem pitting in particular. For instance, the *V. riparia* × *V. rupestris* crosses were much more susceptible to fanleaf than *V. riparia* × *V. berlandieri* crosses (Egger et al. 1985). This might be one reason that SO4 seemed like such a superior rootstock for a time in Europe. However, SO4 is very sensitive to Mg deficiency and prone to thyllosis (where the sap becomes blocked and sudden leaf wilting occurs). Today SO4 is being replaced in some areas with clean rootstock of some of the older, common rootstocks, including 3309 C, 101–14 Mgt, and Riparia Gloire (Boulay 1982). The University of California at Davis has released several *V. vinifera* × *V. rotundifolia* hybrid rootstocks reportedly highly tolerant to fan-leaf virus vectors (Morton and Jackson 1988).

Grapevine rootstocks with resistance to nematode vectors of viruses have been shown by Oliver and Fuchs (2011) (Table 5.11).

5.4   Crown Gall *(Agrobacterium Tumefaciens)*

Crown gall is a grapevine disease often encountered in vineyards prone to winter cold injury (Filo et al. 2013). Lehoczky (1968) hypothesized that the pathogen overwinters primarily in the root system and under moist conditions in the spring, root pressure causes xylem fluids to sweep bacterial cells from the roots upward, where they may be attracted to wounds.

Agrobacterium strains have been divided into three biovars, based on physiological and biochemical properties: biovar I strain, *A. tumefaciens* C58; biovar II strain *A. radiobacter* K84; and biovar III strain *A. vitis* S4 (Slater et al. 2009, Lacroix and Citovsky 2013). Crown gall is neoplastic growth (tumor) resulting from uncontrolled cell proliferation induced by *Agrobacterium tumefaciens*, usually at the interface between the root and the stem of the host plant (Lacroix and Citovsky 2013). Production of crown gall-free vines (rootstock and cultivars) is important to reduce the impact of winter injury (Ker et al. 2007).

Vitis vinifera are highly susceptible to crown gall (Szegedi et al., 1984). Several studies have shown that some rootstocks are resistant to crown gall (Ferreira and van Zyl 1986; Goodman et al. 1993; Heil 1993; Stover 1993; Süle et al. 1994).

Susceptibility of 45 rootstocks to five *A. tumefaciens* strains (1771, 2160, 2164, 2221, and W7) was determined by Ferreira and van Zyl (1986). Paulsen 775, 3309 C, Kober 125AA, Freedom, and Harmony showed minimal tumor growth. Paulsen 775 was immune to all of the strains tested. Solonis, 101–14 Mgt., Berlandieri 13/5, and Kober 5BB appeared to be resistant. The Jacquez, US 2–1, US 24–23, Muscat d' Alexandrie, and US 24–10 rootstocks were the most susceptible. Host range differences occurred among the five strains of *A. tumefaciens* on the different rootstocks and five different host range patterns were evident. *A. tumefaciens* biovar three strain 1771 was the most virulent, and strain 2221 was the least virulent of the strains used (Ferreira and van Zyl 1986).

Table 5.11 Grapevine rootstocks with resistance to dagger nematode (*Xiphinema spp.*) vectors of viruses, based on yield, vigor determination, or other measures of host performance (root damage)

Resistance to *Xiphinema americanum*	Resistance to *Xiphinema index*
O39-16	1045 Paulsen
	110 Richter[a, d]
	1202 C
	1613 C[a, d]
	161–49 C[a]
	26G[d]
	3309 C[d]
	57 Richter[a]
	770 Paulsen[a]
	Börner
	Dog Ridge[b, d]
	Freedom
	Gloire de Montpellier[a, d]
	Grande Glabre[a]
	Harmony[c,d]
	O39–16
	O43–43[a]
	Salt creck[b, d]
	Schwarzmann[a]
	SO4[d]
	Teleki 5A
	Teleki 5C[d]

[a] Moderate resistance, [b] Resistance to California *X. index* populations only, not other *X. index* populations tested, [c] Resistance to Italian, Israeli, and French *X. index* populations only, not to a Californian *X. index* population, [d] Resistance not confirmed in some reports
Modified from Oliver and Fuchs (2011) with permission from ASEV, © 2011 American Society for Enology and Viticulture. AJEV 62:438–451

Relative levels of crown gall susceptibility determined in 43 *Vitis* genotypes by Stover et al. (1997). No genotype was immune to crown gall. Rootstocks 3309 C, 5C, RGM, and 101–14 Mgt, were among the most crown gall resistant genotypes, and 110 R, 420 A, and Dogridge were highly susceptible to crown gall.

Later in a study, grape rootstocks were evaluated by some bacterial strains for biological control of grape crown gall (Burr et al. 1998). Biological control involves antagonistic organisms, which are applied to grapevine roots to reduce the concentration of pathogenic Agrobacterium strains (Filo et al. 2013).

Table 5.13 Phytophthora resistant attributes of the most popular rootstocks in South Africa (Southey 1992; Hunter et al. 2016; van Schalkwyk 2017)

Rootstock	Phytophthora
99 R	Susceptible
101–14 Mgt	Moderately resistant
110 R	Moderately susceptible
140 Ru	Moderately susceptible
143-B Mgt	Resistant
420-A Mgt	Moderately resistant
775 P	Resistant
1045 P	Moderately resistant
1103 P	Susceptible
3306 C	Susceptible
3309 C	Moderately susceptible
Jacquez	Resistant
Ramsey	Very resistant-Resistant
Rupestris du Lot	Susceptible
SO4	Susceptible
USVIT 8–7	Very resistant-Resistant

5.5 Phytophthora

A cool, rainy spring weather initiates rapid reproduction of the fungal spores of phytophthora. Moist soil conditions and warm temperatures lead to the rapid development of phytophthora crown, and root rot, especially when grapes are cultivated on their own-roots. *P. cinnamomi*, *P. cryptogea*, and *P. drechsleri* were isolated from diseased root or crown tissue of the table grape cultivars by Latorre et al. (1997) in Chile. Different popular rootstocks were evaluated for phytophthora resistant attributes in South Africa where143B Mgt, Jacquez, and USVIT 8–7 showed very good resistance (van Schalkwyk 2017) (Table 5.13).

5.6 Grapevine Trunk Disease (GTD)

Latent grapevine trunk disease infection in rootstock cuttings is a major source of GTDs in grafted vines and thus new vineyards. Multiple pruning cuts at the crown predispose mother vines to GTD infection. The pruning cuts can make the infection via the xylem and also by direct penetration of the bark. Controls are limited and not always fully effective against all known GTD pathogens (*Diatrypaceae* spp., *Botryosphaeriaceae* spp., *Basidiomycetes*, *Phaeoacremonium* spp., *Pa. chlamydospore*, *Cadophora luteo-olivacea*, Black-foot pathogens), especially in wet conditions (Gramaje et al. 2017).

References

Anders F (1958) Aminosauren als gallenerregende Stoffe der Reblaus (*Viteus* [Phylloxera] *vitifoloii* Shimer). Experientia 14:62–63

Anwar SA, McKenry MV, Kaku S (1999) Resistance of ten grape rootstocks against six *Meloidogyne* spp. J Nematol 31:522

Anwar SA, McKenry M, Ramming D (2002) A search for more durable grape rootstock resistance to root-knot nematode. Am J Enol Vitic 53(1):19–23

Börner C (1943) Die ersten reblausimmunen Rebenkreuzungen. Angew Bot 25:126–143

Boubals D (1966) A study of the distribution and causes of resistance to radicicolous phylloxera in the Vitaceae. Annales D'amelioration Des Plantes 16(2):145–183

Boulay H (1982) Absorption différenciée des cépages et des portegreffes en Languedoc. Prog Agric Vitic 99:431–434

Burr TJ, Bazzi C, Süle S, Otten L (1998) Crown gall of grape: biology of Agrobacterium vitis and the development of disease control strategies. Plant Dis 82:1288–1297

Chitambar JJ, Raski DJ (1984) Reactions of grape rootstocks to Pratylenchus vulnus and *Meloidogyne* spp. J Nematol 16:166–170

Clingeleffer P, Smith B (2011) Rootstock breeding and development for Australian wine grapes, CSIRO plant industry, final report to grape and wine research & development corporation. CSP 05/03, p 102

Corrie AM, Hoffmann AA (2004) Fine-scale genetic structure of grape phylloxera from the roots and leaves of Vitis. Heredity 92(2):118–127

Corrie AM, Kellow A, Buchanan G, Van Heeswijck R (1998) Phylloxera biotypes in Australia. Aust Grapegrower Winemaker 417:32–36

Corrie AM, van Heeswijck R, Hoffmann AA (2001) Life cycle of grape phylloxera: management implications. Aust. Grapegrower Winemaker 450:13–16

Corrie AM, Crozier RH, van Heeswijck R, Hoffmann AA (2002) Clonal reproduction and population genetic structure of grape phylloxera, Daktulosphaira vitifoliae, in Australia. Heredity 88:203–211

Dry N (2005) Final report to grape and wine research and development corporation, phylloxera and grape industry board of South Australia. PGI 05/01, p 22

Dry N (2008) Grapevine rootstocks. Selection and management for South Australia vineyards. Phylloxera and grape industry board of South Australia. Lythrum Press, p 85

Egger E, Borgo M, Antoniazzi P (1985) Tolerance of grapevine rootstocks to some virus or virus-like diseases. Rivista Di Viticoltura e Di Enologia 38(5):302–307

Ferreira JHS, van Zylb FGH (1986) Susceptibility of grape-vine rootstocks to strains of agrobacterium tumefaciens biovar 3. S Afr J Enol Vitic 7(2):101–104

Ferris H, Zheng L, Walker MA (2012) Resistance of grape rootstocks to plant-parasitic nematodes. J Nematol 44(4):377–386

Filo A, Sabbatini P, Sundin GW, Zabadal TJ, Safe GR, Cousins PS (2013) Grapevine crown gall suppression using biological control and genetic engineering: a review of recent research. Am J Enol Vitic 64(1):1–14

Fuchs M (2020) Grapevine viruses: a multitude of diverse species with simple but overall poorly adopted management solutions in the vineyard. J Plant Pathol 102:643–653

Goodman RN, Grimm R, Frank M (1993) The influence of grape rootstocks on the crown gall infection process and tumor development. Am J Enol Vitic 44:22–26

Gramaje D, Waite H, Armengol J, Billones-Baaijens R, Halleen F, Di Marco S, Zekri O, Smart R (2017) A protocol proposal for the management of grapevines to reduce latent GTDs infections in cuttings. http://managtd.eu/images/uploads/content/258/WG4_Bologna%202017_DGramaje_FA1303.pdf

Gu S, Ramming D (2005a) Viticultural performance of Syrah grapevines on new USDA-ARS rootstocks for winegrape production in the San Joaquin Valley of California. 56th annual meeting of American society for enology and viticulture. Seattle, Washington. Am J Enol Vitic 56:312A

Gu S, Ramming D (2005b) Viticultural performance of Thompson Seedless grapevines on new USDA-ARS rootstocks for raisin production in the San Joaquin Valley of California. 56th annual meeting of American society for enology and viticulture. Seattle, Washington. Am J Enol Vitic 56:312A

Heil M (1993) Untersuchungen zur Resistenz von Vitis gegen Agrobacterium tumefaciens. Hohenheim, Germany: University of Hohenheim, PhD thesis

Hunter JJ, Archer E, Van Schalkwyk D, Strever AE, Volschenk CG (2016) Grapevine roots: interaction with natural factors and agronomic practices. Acta Hortic 1136:63–80

UC IPM (2021) UC pest management guidelines: grapes. UC ANR Publication 3448, Retrieved 5 March 2019. http://ipm.ucanr.edu/PMG/r302200111.html

Kellow AV, Sedgley M, Sedgley R (2004) Interaction between Vitis vinifera and grape phylloxera: changes in root tissue during nodosity formation. Ann Bot 93(5):581–590

Kellow AV (2000) A study of the interaction between susceptible and resistant grapevines and phylloxera, Department of Horticulture, Viticulture and Oenology, Adelaide University, PhD. dissertation, p 298

Ker KW, Slingerland K, Fraser H, Fisher KH (2007) Ontario's experience with wind machines for winter injury protection of grapevines and tender fruit. pp 75–78. In: Proceedings "Understanding and preventing freeze damage in vineyards" Columbia, MO. 5–6 Dec 2007. p 107

Kerbs SJ (1995) The effect of phylloxera biotype and water stress on phylloxera resistance in grape rootstocks. PhD. dissertation, University of California, Davis

Lacroix B, Citovsky V (2013) Agrobacterium. In: Maloy S, Hughes K (eds) Brenner's encyclopedia of genetics, vol 1, 2nd edn. Academic Press, San Diego, pp 52–54

Laliman L (1869) Nouvelle phase du phylloxera decouverte, fin juillet, de la forme vivant dans les galles des feuilles, Bulletin of the Society of Agriculture and Horticulture of Vaucluse, p 254

Latorre BA, Wilcox WF, Bañados MP (1997) Crown and root rots of table grapes caused by Phytophthora spp. Chile, Vitis 36(4):195–197

Lehoczky J (1968) Spread of Agrobacterium tumefaciens in the vessels of the grapevine after natural infection. Phytopathol Z 63:239–246

Lund K, Riaz S, Walker MA (2017) Population structure, diversity and reproductive mode of the grape phylloxera (Daktulosphaira vitifoliae) across its native range, e. PLoS ONE 12(1):e0170678. https://doi.org/10.1371/journal.pone.017067

Maeda K, Kubuchi T, Kasajima I, Li C, Yamagishi N, Yamashita H, Yoshikawa N (2020) Virus-induced flowering by apple latent spherical virus vector: effective use to accelerate breeding of grapevine. Viruses. https://doi.org/10.3390/v12-10070

Martelli GP, Adams MJ, Kreuze JF, Dolia VV (2007) Family Flexiviridae: a case study in virion and genome plasticity. Ann Rev Phytopathol, Palo Alto 45:73–100

Martelli GP (2018) Where grapevine virology is heading to. In: Proceedings of the 19th congress of the international council for the study of virus and virus-like diseases of the grapevine (ICVG). Santiago, Chile, 9–12 April 2018

McKenry MV, Kretsch JO, Anwar SA (2001) Interaction of selected rootstocks with ectoparasitic nematodes. Am J Enol Vitic 52:304–309

McKenry MV, Kretsch JO, Anwar SA (2001) Interaction of selected rootstocks with ectoparasitic nematodes. Am J Enol Vitic 52:310–316

Morton LT, Jackson LE (Jan 1988) Myth of the universal rootstock: the fads and facts of rootstock selection. In: Proceedings second international cool climate viticulture and oenology symposium. Auckland, New Zealand

Nicol JM, Stirling GR, Rose BJ, May P, Van Heeswijck R (1999) Impact of nematodes on grapevine growth and productivity: current knowledge and future directions, with special reference to Australian viticulture. Aust J Grape Wine Res 5(3):109–127

Niklowitz W (1955) Histologische studien an reblausgallen und reblausabwehrnekrosen. Phytopathol Z 24:299–340

Oliver JE, Fuchs M (2011) Tolerance and resistance to viruses and their vectors in Vitis sp.: a virologist's perspective of the Literature. Am J Enol Vitic 62(4):438–451

Pàstena B (1976) Trattato di Viticoltura Italian. Edagricola, Bologna, pp 379–403

Pavloušek P (2012) Screening of rootstock hybrids with *Vitis cinerea* Arnold for phylloxera resistance. Cent Eur J Biol 7(4):708–719. https://doi.org/10.2478/s11535-012-0062-z

Porten M, Schmid V, Rühl EH (2000) Current problems with phylloxera on grafted vines in Germany and ways to fight them. In: Powell KS, Whiting J (eds). Proceeding of the international symposium on grapevine phylloxera management. Organized by agriculture Victoria-Rutherglen. Melbourne, pp 89–98

Powell KS (2009) Final report to: grape and wine research & development corporation. Project number: DPI 06/02. DPI-Rutherglen p 107

Rahemi A (2016) Breeding of grapevine rootstocks for adaptability to the environmental stresses, third national conference of grape and raisin conducted by research institute of grape and raisin (RIGR), 27–29 September 2016. Malayer University, Malayer, Iran

van Schalkwyk D (2017) Fruit production: get the right rootstock for planting grapes, Dec14, 2017, Retrieved: 9 Sep 2019. https://www.africanfarming.com/fruit-production-get-right-rootstock-planting-grapes

Schmid J, Manty F, Rühl EH (2015) Optimisation of sustainability of grapevine varieties by selecting rootstock varieties under different environmental conditions and creating new rootstock varieties. Retrieved 5 Oct 2019. http://www.viticulturasustentavel.com/pdfs/Prof_Joachim_Schmid.pdf

Slater SC, Goldman BS, Goodner B, Setubal JC, Farrand SK, Nester EW, Burr TJ, Banta L, Dickerman AW, Paulsen I, Otten L, Suen G, Welch R, Almeida NF, Arnold F, Burton OT, Du Z, Ewing A, Godsy E, Heisel S, Houmiel KL, Jhaveri J, Lu J, Miller NM, Norton S, Chen Q, Phoolcharoen W, Ohlin V, Ondrusek D, Pride N, Stricklin SL, Sun J, Wheeler C, Wilson L, Zhu H, Wood DW (2009) Genome sequences of three agrobacterium biovars help elucidate the evolution of multichromosome genomes in bacteria. J Bacteriol 191(8):2501–2511

Southey JM (1992) Grapevine rootstock performance under diverse conditions in South Africa. Paper presented at: rootstock seminar: a Worldwide Perspective, Reno, Nevada, USA

Stamp JA (2011) Nematodes and viticulture, Poor vineyard performance? Consider nematodes as likely participants, Wine Business Monthly, March 2011 Issue, Retrieved 5 March 2019. https://www.winebusiness.com/wbm/index.cfm?go=getArticle&dataId=86294

Stover EW, Swarz HJ, Burr TJ (1997) Crown gall formation in a diverse collection of *Vitis* genotypes inoculated with *Agrobacterium vitis*. J Am Soc Enol Vitic 48:26–32

Stover EW (1993) Resistance to crown gall in Vitis: studies directed toward the identification of crown gall resistant rootstocks. College Park, USA University of Maryland, PhD thesis

Süle S, Mozsár J, Burr TJ (1994) Crown gall resistance of *Vitis* spp. and grapevine rootstocks. Phytopathology 84:607–611

Szegedi E, Korbuly J, Koleda I (1984) Crown gall resistance in East-Asian *Vitis* species and their *V. vinifera* hybrids. Vitis 23:21–26

Viala P, Ravaz L (1903) American vines: their adaptation, culture, grafting and propagation. R Dubois and E.H. Twight, translators. Freygang-Leary Co., San Francisco. p 299

Progressive Viticulture (2011) Selecting a rootstock for a wine grape vineyard. http://www.progressivevit.com/selecting-a-rootstock-for-a-wine-grape-vineyard/. (Accessed Dec 2016)

Vorwerk S, Forneck A (2006) Reproductive mode of grape phylloxera (*Daktulosphaira vitifoliae*, Homoptera: Phylloxeridae) in Europe: molecular evidence for predominantly asexual populations and a lack of gene flow between them. Genome 49:678–688

Walker MA (1991) Characteristics of *Vitis* species commonly used in rootstock breeding. ASEV Publication, Davis, CA, Alternative Rootstock Update, pp 1–7

Walker MA, Riaz S, Fort KP, Heinitz CC, Uretsky J, Pap D (2018) The history of grape rootstocks and how future needs might be addressed. Department of viticulture and enology, University of California, Davis, CA, USA, http://www.acenologia.com. Retrieved 22 Dec 2018

Whiting JR (2003) Selection of grapevine rootstocks and clones for Greater Victoria. Department of Primary Industries: Victoria, p 40

Whiting JR (2012) Rootstock breeding and associated R&D in the viticulture and wine industry, Australian Government, Grape and Wine Research and Development Corporation (GWRDC). https://gwrdc.com.au

Chapter 6
Abiotic Stresses and Rootstock Selection

6.1 Soil Conditions and Problems

6.1.1 Soil Types

Choosing grape rootstocks that are suited to different soil type was determined by Shaffer et al. (2004) (Table 6.1).

Similarly, rootstocks with some degree of tolerance to soil limitations (water stress, wet soil, saline soil, sodium and chloride, acid soils, and alkaline soils) were summarized by Progressive Viticulture (2011) (Table 6.2).

6.1.2 Soil Texture and Soil Depth

The wetting pattern of soil is affected by soil texture. In clay soils, the wetting area is shallower, while in sandy soil, the wetting area is deeper (Dry 2015) (Fig. 6.1). Root distribution patterns and geotropic angles vary among rootstocks, and there is a need to select the appropriate rootstock for each soil type. Sweet and Walker (2007) proposed that the geotropic angles for the roots of grape rootstocks be classified into three classes: (1) shallow roots with 0 to 35-degree angles, (2) medium roots with 36 to 54-degree angles, and (3) deep roots with 55 to 90-degree angles. *Vitis riparia* usually has shallow roots with wide angles, while *Vitis rupestris* has deeper roots with closer angles. *Vitis riparia* (Riparia Gloire) has been shown to have a shallower root system with roots having higher angles of distribution (10 to 15-degree angles) compared with 70 degrees in Rupestris du Lot (Guillon 1905). *V. riparia* has lower root weights than *V. rupestris* (Branas and Vergnes 1957) and was the only species to produce less root material than shoot material (Ravas 1902). *Vitis berlandieri*

The original version of the chapter was revised: Figure 6.5 has been removed. The correction to the chapter is available at https://doi.org/10.1007/978-3-030-99407-5_12

63

A. Rahemi et al., *Grape Rootstocks and Related Species*, https://doi.org/10.1007/978-3-030-99407-5_6

Table 6.1 Adaptibility grape rootstocks based on soil type and soil depth (Shaffer et al. 2004 derived from Howell 1987; Pongrácz 1983; Galet 1998; Jackson 2000; Taylor 2002; and Shaffer 2002)

	Shallow	Deep	
	Clay	Silt or loam	Sandy
RGM	Low	Low	Moderate low
3309 C	Moderate high	Low	Moderate low
101–14 Mgt	Moderate low	Low	Moderate low
Schwarzmann	Low	Moderate low	Moderate high
420A Mgt	Moderate high	Low	Moderate low
5BB	Moderate low	Low	Low
SO4	Low	Low	Low
5 C	Moderate high	Moderate high	Low
125 AA	Moderate high	Moderate low	Low
161–49 C	Low	Low	Moderate low
99 R	Moderate low	Moderate high	Moderate low
110 R	High	Moderate high	Moderate high
1103 P	Moderate high	Moderate high	Moderate high
140 Ru	Moderate high	Moderate high	High
1616 C	Low	Moderate low	Moderate low
44–53 M	Moderate low	Moderate high	Moderate low
Gravesac	Moderate low	Moderate high	Moderate high

Table 6.2 Rootstocks with some degree of tolerance to soil limitations (Progressive Viticulture 2011)

Water stress	Wet soils	Saline soil	Sodium and Chloride	Acid soils	Alkaline soils
99 R	Riparia Gloire	140 Ru	140 Ru	99 R	5BB
110 R	SO4	Schwarzmann	Schwarzmann	140 Ru	420A
140 Ru	Schwarzmann	1616 C	1616 C		140 Ru
1103 P	1616 C	Harmony	Ramsey		
SO4	Harmony	Ramsey			
5BB		Dogridge			
125AA					
Freedom					
St. George					
Ramsey					
Dogridge					
1616 C					

(Copyright © 2011 Progressive Viticulture), http://www.progressivevit.com/selecting-a-rootstock-for-a-wine-grape-vineyard/

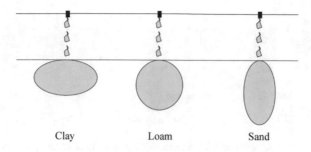

Fig. 6.1 Soil texture influence on the wetting pattern (Dry 2015)

has been shown to have an intermediate distribution pattern (Perold 1927; Schmid et al. 2015; Rahemi 2016) (Fig. 6.2). However, Sweet and Walker (2007) reported that some genotypes differ significantly in root angle from one or from the parents. Using this information in combination, Dry (2015) presented a hypothetical match of root distribution patterns of interspecific hybrids with different soil types (Figs. 6.3 and 6.4). For choosing the rootstocks also it is important to consider the depth of the soil. In shallow depth soil, it would be better to choose rootstocks with *V. riparia* background, and in deeper depth soil, it would be better to choose rootstocks with *V. berlandieri* and *V. rupestris* background (Rahemi 2016). Winkler et al. (1974), determined soil depth categories for vineyards. They mentioned that the soils with less than 75 cm deep should be altered by ripping, deep plowing, etc., before planting.

Pavloušek (2013) presented rootstocks recommendations based on drought tolerance and pedological conditions of the site (Table 6.3):

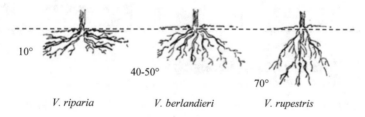

Fig. 6.2 Rootstock root distribution patterns and geotropic angles (Perold 1927; Schmid et al. 2015; modified by Rahemi 2016)

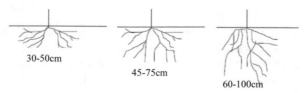

V. riparia × V. berlandieri   V. riparia × V. rupestris   V. berlandieri × V. rupestris

Fig. 6.3 Hypothetical rootstock root distribution patterns (Dry 2015)

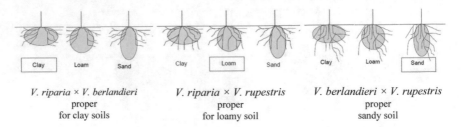

V. riparia × V. berlandieri	V. riparia × V. rupestris	V. berlandieri × V. rupestris
proper	proper	proper
for clay soils	for loamy soil	sandy soil

Fig. 6.4 Choosing grape rootstocks hypothetically based on their genetic, root distribution, and soil wetting pattern (Dry, 2015; modified by Rahemi 2016)

Table 6.3 Recommended grape rootstocks based on their drought tolerance and pedological conditions (Pavloušek 2013)

Soil profile characteristics	Vineyard water status	Recommended rootstocks
Soil depth <20 cm: sand, loam, or clay, including any root-impeding subsoil	Rainfed	110 R, 140 Ru, 1103 P
	Irrigated	110 R, 140 Ru, 1103 P, Ramsey
Soil depth 20–75 cm, sands, loams, or clays, with no root-impeding subsoil	Rainfed	99 R, 110 R, 140 Ru, 1103 P, Ramsey, Kober 5 BB
	Irrigated	99 R, 110 R, Ramsey, Kober 5BB, Teleki 5C, Schwarzmann, SO4, 420A, 101–14 Mgt (loams and clays)
Soil depth >75 cm, uniform or gradational profile of sand, loam, or clay	Rainfed	99 R, 110 R, 1103 P, Ramsey (in sand), Kober 5BB
	Irrigated	SO4, 101–14 Mgt, Teleki 5C, Schwarzmann, 3306 C, 3309 C, 420A

6.1.3 Soil Water Deficit (Drought)

Grapevines react in a number of different ways to tolerate water constraints. Even though grapevines are considered relatively tolerant to water deficits, grapevine growth and yield can be seriously reduced under water deficit. Drought-tolerant rootstocks are expected to enable the scion to grow and yield when the water supply is limited. Rootstocks have been shown to control water extraction capacity and scion transpiration (Serra et al. 2014).

The roots of different rootstocks have morphological and physiological differences that result in different abilities to take up water and nutrients. As indicated above, roots of different rootstocks have different distributions allowing them to be distributed in different soil zones (Harmon and Snyder 1934; Champagnol 1984), and deeper roots give them access to a greater moisture supply. Stomatal characteristics of the leaves have been shown to be related to rootstock type. On *V. riparia* × *V. rupestris* hybrids, stomatal densities and size, and rates of stomatal conductance, are higher than in *V. rupestris* × *V. berlandieri* or *V. riparia* × *V. rupestris* hybrids (Scienza and Boselli 1981). Rootstocks can impart drought resistance to the leaves of their scions (Carbonneau 1985; Morton and Jackson 1988).

Cox (2014) has mentioned that the rootstocks supply the grafted vine with a root system. The root system as a result of the choice of rootstock affects the vertical depth and horizontal spread of the roots along with the root density. These genetic differences also influence the drought tolerance of a grafted vine. For example, a rootstock's ability to cope with dry conditions may be attributed to an ability to explore large volumes of soil with thick root systems.

Drought tolerance is scion and site-specific, and absolute drought tolerance is, therefore, difficult to classify. However, in general, rootstocks with a greater proportion of fine shallow roots with a diameter <2 mm will be more prone to drought and be lower in vigor than a rootstock with thicker roots and deeper root systems (Soar and Loveys 2006). Dry (2015) classified 13 grape rootstocks for drought tolerance as being highly tolerant, tolerant, moderately susceptible, and susceptible. Ramsey and 1103 P were classified as having high tolerance (Table 6.4).

Southey (1992), Hunter et al. (2016), and van Schalkwyk (2017), also compared the most popular rootstocks in Southern South Africa to drought conditions. He separated them into excellent, good, fair, moderate-poor, and poor classifications (Table 6.5).

The drought tolerance of the 34 common European rootstocks was assessed by Pavloušek (2013) as being very low, low, medium, high, and very high. The rootstocks 1103 Paulsen and 775 Paulsen had high to very high tolerance (Table 6.6).

Serra et al. (2014) reviewed 16 grape rootstocks based on a classification of drought adaptation proposed by either Samson and Castéran (1971) (good, average,

Table 6.4 Classification drought tolerance (Dry 2015)

Classification	Rootstock
Highly tolerant	Ramsey
	1103 P
Tolerant	140 Ru
	110 R
	99 R
Moderately susceptible	5BB Kober
	5C Teleki
	SO4
Susceptible	101–14 Mgt
	Schwarzmann
	3309 C
	Riparia Gloire
	420A

Table 6.5 Drought resistant attributes of the most popular rootstocks in South Africa (Southey 1992; Hunter et al. 2016; van Schalkwyk 2017)

Rootstock	Drought resistance
99 R	Good
101–14 Mgt	Fair
110 R	Excellent
140 Ru	Excellent
143-B Mgt	Good
420-A Mgt	Good
775 P	Excellent
1045 P	Good
1103 P	Excellent
3306 C	Fair
3309 C	Poor
Jacquez	Fair
Ramsey	Moderately poor
Rupestris du Lot	Good
SO4	Poor
USVIT 8–7	Fair

weak, and bad) or Fregoni (1977) (highly resistance, mid resistance, average resistance, and weak resistance) or Carbonneau (1985) (high resistance, resistance, sensitive and very sensitive). The rootstocks 110 R and 44–53 M showed the best resistance in all three classifications With 140 Ru and 1103 P having the best resistance in at least in two reports (Table 6.7).

6.1.4   Excess Soil Water (Saturation or Bad Drainage)

All vines perform poorly in flooded soil conditions. Grapevine symptoms of waterlogging include desiccation of the shoot apex, flagging of leaves, necrotic areas on leaves, senescence of basal leaves, and regeneration of roots near the water surface (Striegler et al. 1993). Ungrafted vines usually show better tolerance, and rootstocks are generally more susceptible to waterlogging than own-rooted *V. vinifera* (Dry 2015). *V. cinerea* is considered to be a waterlogged tolerant species (Pongrácz 1983). Striegler et al. (1993) compared St. George, 3309 C, Riparia Gloire, Kober 5BB, Seyval (own-root), and Cynthiana (*Vitis aestivalis*) to soil flooding conditions under greenhouse conditions. In general, the rate of shoot elongation was the most sensitive, and water use efficiency was the least sensitive parameter to waterlogging. St. George, 3309 C, and Riparia Gloire showed the most tolerance, while Kober 5BB, Seyval, and Cynthiana showed the most sensitivity to waterlogging. Dry (2015) reported that the rootstocks 99 R, 420A, 140 Ru, and Ramsey should not be used in

Table 6.6 Drought tolerance of the most common European rootstocks (Pavloušek 2013)

Rootstock	Drought tolerance
161–49 C	+
RGM	+
5C	+(+)
101–14 Mgt	+(+)
Rupestris du Lot	++
Fercal	++
125 AA	++ (+)
Schwarzmann	++ (+)
5 BB	+++
R.S.B.1	+++
3309 C	+++
Cosmo 2	+++
Cosmo 10	+++
Sori	+++
1616 C	+++
26 G	+++
SO4	+++(+)
Binova	+++(+)
Teleki 8B	+++(+)
Börner	+++(+)
Rici	+++(+)
Cina	+++(+)
Golia	+++(+)
420A	++++
140 Ru	++++
Richter 110	++++
Richter 99	++++
Gravesac	++++
Sorisil	++++
41B Mgt	++++
333 E.M	++++
Georgikon 28	++++
1103 Paulsen	++++(+)
775 Paulsen	++++(+)

Drought tolerance scales (+ = Very low, ++ = Low, +++ = Medium, ++++ = High, +++++ = Very high)

Table 6.7 Rootstock classification based on adaptation to drought, as proposed by Samson and Castéran (1971), Fregoni (1977), and Carbonneau (1985) and derived from Serra et al. (2014)

Name	Samson and Castéran	Fregoni	Carbonneau
110 R	Good	High resistance	High resistance
140 Ru	Average	High resistance	High resistance
44–53 M	Good	High resistance	High resistance
1103 P	Good	High resistance	Resistance
SO4	Weak	Weak resistance	Resistance
99 R	Average	Average resistance	Resistance
3309 C	Good	Weak resistance	Sensitive
420A Mgt	Weak	Weak resistance	Sensitive
Fercal	Average	–	Sensitive
5BB	Bad	Weak resistance	Sensitive
161–49 C	Weak	Mid resistance	Sensitive
41B Mgt	Average	High resistance	Sensitive
Rupestris du Lot	Bad	Weak resistance	Sensitive
101–14 Mgt	Bad	Weak resistance	Very sensitive
Riparia Gloire de montpellier	Bad	Weak resistance	Very sensitive
333EM	Good	Mid resistance	Very sensitive

waterlogged soils since they are susceptible, while Schwarzmann, SO$_4$, 101–14 Mgt, 1103 P, and 110 R are less susceptible (Table 6.8).

Lambert et al. (2008) evaluated tolerance of 13 rootstocks to excessive moisture caused by poor soil drainage, including Teleki 5C, Kober 5BB, 3309 C, 101–14 Mgt, Richter 110, Paulsen 1103, 420A, 1616 C, St. George, Malègue 44–53, Ramsey, Harmony, and Freedom. They categorized them as being low, medium, high, very high, or variable. 1616 C, 44–53 M, and 1103 P were determined as having high tolerance to excessive soil moisture (Table 6.9).

Table 6.8 Grape rootstock susceptibility to bad drainage (Dry 2015)

Susceptible	Less susceptible
99 R	Schwarzmann
Ramsey	SO4
140 Ru	110 R
420A	1103 P
	101–14 Mgt*

* Susceptible in early years, but more tolerant as the vine develops

Table 6.9 Grape rootstock tolerance to excessive moisture caused by poor drainage (Pongrácz 1983; Christensen 2003; Lambert et al. 2008)

Rootstock	Tolerance to poor drainage (wet feet)
St. George	L/M
1616 C	H
3309 C	L/M
44–53 M	H
101–14 Mgt	M/H
420A	L/M
5BB	Var
5C	Var
1103 P	H
110 R	Var
Freedom	L
Harmony	L
Ramsey	L/M

L = low, M = medium, H = high, VH = very high, and Var = variable

The most popular rootstocks in South Africa has been classified to excellent, good, fair, moderate and poor resistant to soil wet (Southey 1992; Hunter et al. 2016; van Schalkwyk 2017). (Table 6.10).

6.1.5 Soil Acidity (pH)

Soils below pH 5.5, generally lead to nutrient deficiency (N, P, K, S, Ca, and Mg) or nutrient toxicity in grapevines (Al, Cu, and Mn). Dry (2015) recommended 140 Ru, 1103 Paulsen, 99 Richter and 110 Richter for use on soils with low pH (Table 6.11).

Table 6.10 Soil wet resistant attributes of the most popular rootstocks in South Africa (Southey 1992; Hunter et al. 2016; van Schalkwyk 2017)

Rootstock	Wet
99 R	Poor-Fair
101–14 Mgt	Excellent
110 R	Good
140 Ru	Fair
143-B Mgt	Excellent
420-A Mgt	Fair
775 P	Good
1045 P	Good
1103 P	Good
3306 C	Good
3309 C	Fair
Jacquez	Good
Ramsey	Moderate
Rupestris du lot	Good
SO4	Good
USVIT 8–7	Good–Excellent

Table 6.11 Soil acidity (Dry 2015)

Recommended rootstocks	Rootstock to avoid
140 Ru	Schwarzmann
1103 P	101–14 Mgt
99 R	SO4
110 R	Ramsey
	5 C Teleki
	5 BB Kober
	3309 C

6.1.6 Soil Alkalinity (Lime Content)

Vitis vinifera reveals symptoms of Fe deficiency when grafted on some lime-susceptible rootstocks, even though the response varies as a function of the grape variety. Grapevine rootstocks adopt "strategy I" to overcome iron chlorosis, increasing root iron uptake, proton extrusion, and reducing capacity (Bavaresco and Poni 2003). The responses of susceptible grapevines to lime-stress conditions are different and include a shoot and root growth reduction, yield reduction (Ryan et al. 2013), and typical leaf interveinal yellowing (Bavaresco and Poni 2003). In calcareous soils, the carbonates are mainly represented by calcite ($CaCO_3$) and dolomite [$CaMg(CO_3)$]. The total carbonate content of the soil is usually expressed as calcium carbonate equivalent (CCE). Active calcium carbonate equivalent (ACCE) is

	CPI value	Intensity of chlorosis
Table 6.12 The chlorosis intensity degrees for different CPI values (Rusu et al. 2005; Lupascu et al. 2009)	0	None
	≤ 5	Low
	6–15	Medium
	16–35	High
	≥ 36	Very high

one method (Drouineau 1942) that is systematically used for categorizing grape rootstocks to soil alkalinity (Champagnol 1984). This method is sometimes refined to a chlorotic power index (CPI or IPC), which is the ratio between ACCE and neutral NH_4 oxalate-extractable Fe (as a predictor of the Fe chlorosis risk in grapevines) (Juste and Pouget 1972). The chlorotic power index is a measure of the rootstocks' adaptability to tolerate lime-induced chlorosis, also known as iron deficiency chlorosis. CPI is defined as the ratio between active lime and assimilable iron and is calculated as (active $CaCO_3$, %/(Fe, ppm)^2) $\times 10^4$. The chlorosis intensity degrees for different CPI values defined by Lupascu et al. (2009) (Table 6.12).

The value of the Chlorotic Power Index (CPI) was determined for grape rootstocks (Pouget and Delas 1989; Hamman et al. 1998) (Table 6.13).

Galet (1998), and Dry (2015) determined 140 Ru, 5BB Kober, 420 A, 110 Richter, 1103 Paulsen, 99 Richter, and SO4 as being the most to least lime tolerant. (Table 6.14).

5C Teleki and Ramsey are not included in Galet's ranking but have been classified in the literature as having lime tolerance.

A test based on the reactivity of soil Fe forms over those based on carbonate-related properties has also been demonstrated in grapevines (de la Torre et al. 2010).

	Rootstock variety	CPI
Table 6.13 Resistance to lime induced chlorosis; a value of the Chlorotic Power Index (CPI) above which symptoms appear (Pouget and Delas 1989; Hamman et al. 1998)	Riparia Gloire	5 low
	101–14 Mgt	10
	3309 C	10
	Gravesac	20
	110 R	30
	1103 P	30
	SO4	30
	420A	40
	161–49 C	50
	41B	60
	140 Ru	90
	Fercal	120 high

Table 6.14 Least to most to least lime tolerant according to Galet (1998) and derived from Dry (2015)

Rootstock	Lime tolerance
SO4	Low
99 R	
1103 P	
110 R	
420 A	
5BB Kober	
140 Ru	High

The chlorosis tolerance of the most common European rootstocks was determined by Hofäcker (2004), and Pavloušek (2013) (Table 6.15).

Active lime tolerance for rootstock varieties has also been shown by Shaffer et al. (2004) (Table 6.16).

Ollat et al. (2016) presented a classification of rootstock tolerance to limestone-induced iron chlorosis. They determined the percent of total lime, percent of active lime, and also the chlorotic power index of 17 commercial grape rootstocks. The most tolerant rootstocks to total lime were Fercal, 140 Ru, 333EM, and 41B, which also showed high CPIs (Table 6.17).

In South Africa, the lime resistant attributes of the most popular rootstocks were studied by Southey (1992), Hunter et al. (2016), and van Schalkwyk (2017) (Table 6.18).

In limestone soils of Germany, Kober 5 BB and other *V. riparia* × *V. berlandieri* hybrids are the preferred rootstocks (Ambrosi and Becker 1978).

Some results from INRA Bordeaux, France, have shown that there is an inverse relationship between lime tolerance and vigor, in that as vigor increases, lime tolerance decreases. This obviously makes the process of breeding difficult when the breeding objective is for high lime tolerance and low conferred vigor (Dry 2005).

6.1.7 Soil Salinity

Excess salinity affects own-rooted vines reducing yields, growth and bunch numbers, and berry weights (Whiting 2004). The rising salinity levels of soils and irrigation waters will restrict root growth, performance, and wine quality of own-rooted vines when levels above 1.8 dSm^{-1} in the soil are reached. Certain rootstocks reduce the uptake of salts, including sodium and chloride (Cox 2014). The performance of grapevine on relatively saline soil was evaluated by Southey and Jooste (1991). They grafted Colombard cultivar onto 25 rootstocks [1045P, 1103 P, 775P, 99 R (clone RY 13), 99 R (clone RY 2), 110 R, 140 Ru, 143-B Mgt, 101–14 Mgt, Ramsey, 13–5 EVEX, Constantia Metallica, Jacquez, Grézot 1, 216-3 Cl and US hybrids, 1–6, 2–1, 3–6, 4–4, 24–10, 24–41, 16–13–23, 16–13–26 12–6–8, and 35–1–15]. The highest

Table 6.15 Chlorosis tolerance of the most common European rootstocks (Hofäcker 2004; Pavloušek 2013)

Rootstock	Chlorosis tolerance
3309 C	+(+)
101–14 Mgt	+(+)
Schwarzmann	++
Rupestris du lot	++
Sori	++
1616 C	++
5C	++(+)
420A	++(+)
Börner	++(+)
Rici	++(+)
Cina	++(+)
5 BB	+++
99 R	+++
Cosmo 2	+++
Cosmo 10	+++
Gravesac	+++
Fercal	+++
41B Mgt	+++
Golia	+++
125 AA	+++(+)
RGM	+++(+)
Sorisil	+++(+)
SO4	++++
Binova	++++
Teleki 8B	++++
140 Ru	++++
1103 P	++++
775 P	++++
110 R	++++
26 G	++++
333 E.M	++++
Georgikon 28	++++
161–49 C	+++++
R.S.B.1	+++++

Chlorosis tolerance scales (+ = Very low, ++ = Low, +++ = Medium, ++++ = High, +++++ = Very high)

Table 6.16 Rootstock varieties, lime tolerance (Shaffer et al. 2004)

Rootstocks	Tolerate to lime
RGM	Low
3309 C	Low
101–14 Mgt	Low
420A Mgt	Moderate high
5BB	Moderate high
SO4	Moderate low
8B	Moderate low
5 C	Moderate low
125 AA	High
161–49 C	Moderate high
99 R	Moderate low
110 R	Moderate low
1103 P	Moderate low
140 Ru	Moderate low
1616 C	Low
44–53 M	Low
Gravesac	Moderate high

Table 6.17 Classification of rootstock tolerance to limestone-induced iron chlorosis. According to Galet (1947), Juste and Pouget (1972), Pouget and Ottenwalter (1978) cited by Champagnol (1984), and derived from Ollat et al. (2016)

Rootstock	Levels of tolerance to lime		
	Total lime%	Active lime%	CPI
Riparia Gloire		6	5
3309 C	15–25	11	10
101–14 Mgt	15–25	9	10
Gravesac			20
Rupestris du lot		14	20
1103 P		17	30
SO4	35	17	30
110 R		17	30
99 R		17	30
5BB		20	40
420A	40–45	17	40
RSB1	40–45	20	50
161–49 C	40–45		60
41B	50–60	40	60
333EM	50–60	40	70
140 Ru	50–60		90
Fercal	>60		120

Table 6.18 Lime resistant attributes of the most popular rootstocks in South Africa (Southey 1992; Hunter et al. 2016; van Schalkwyk 2017)

Rootstock	Lime
99 R	Good
101–14 Mgt	Poor
110 R	Good
140 Ru	Excellent
143-B Mgt	Fair
420-A Mgt	Poor
775 P	Good
1045 P	Good
1103 P	Good
3306 C	Good
3309 C	Good
Jacquez	Good
Ramsey	Moderate
Rupestris du lot	Good
SO4	Fair
USVIT 8–7	Moderate

yields were obtained with 13–5 EVEX, Ramsey, 143-B Mgt, 1045P, and 140 Ru, whereas the US hybrids generally performed poorly, particularly those with Jacquez in their parentage.

Dry (2008) defined salt tolerance of 17 rootstocks (own-roots, 3309 C, 1202 C, K51-40, 5BB Kober, 5C Teleki, 110 R, 99 R, K51-32, 140 Ru, Schwarzmann, 101–14 Mgt, Ramsey, M5489, M6262, M5512, and 1103 P) as being sensitive, moderately sensitive, moderately tolerant and tolerant to soil salinity, where the threshold soil saturation paste salinity was 1.8, 2.5, 3.3, and 5.6 dS m^{-1}, respectively. 1103 P was categorized as being the only salt-tolerant rootstock (Table 6.19).

Table 6.19 Classification of rootstocks based on soil salinity (Tee et al. 2003; Dry 2008)

Classification of soil tolerance	Rootstock	Approximate threshold soil saturation paste salinity (dS/m)
Sensitive	Own-roots, 3309 C, 1202 C, K51-40	1.8
Moderately sensitive	5BB Kober, 5C Teleki, 110 R, 99 R, K51-32	2.5
Moderately tolerant	140 Ru, Schwarzmann, 101–14 Mgt, Ramsey, M5489, M6262, M5512	3.3
Tolerant	1103 P	5.6

Table 6.20 Salinity-resistant attributes of the most popular rootstocks in South Africa (Southey 1992; Hunter et al. 2016; van Schalkwyk 2017)

Rootstock	Salt
99 R	Good
101–14 Mgt	Excellent
110 R	Good
140 Ru	Good
143-B Mgt	Good
420-A Mgt	Poor
775 P	Good
1045 P	Good
1103 P	Fair
3306 C	Poor
3309 C	Poor
Jacquez	Fair
Ramsey	Good
Rupestris du lot	Good
SO4	Poor
USVIT 8–7	Good–Excellent

Salinity-resistant attributes of the most popular rootstocks were studied in South Africa (Southey 1992; Hunter et al. 2016; van Schalkwyk 2017) (Table 6.20). 101–14 Mgt showed an excellent resistance to salinity.

6.1.8 Soil Fertility and Nutrient Deficiency

Clay soils have a greater capacity to store nutrients compared with sand (Gladstones 1992). Consequently, soil fertility generally increases with increasing clay content. The absorbency of certain minerals, e.g., N, K, Ca, Mg, P, varies both among rootstocks and among *V. vinifera* scions (Boulay 1982). N and K are the two most important nutrient elements when selecting and managing rootstocks. Most rootstocks take up and assimilate N more efficiently than own-rooted vines (Dry 2015). Rootstocks significantly affect N concentration at bloom (a factor in fruit set/shatter) but not necessarily at veraison. Rootstocks also influence potassium uptake, and maybe it is an important component of selection criteria depending on site (Dry 2015). High levels of potassium uptake can lead to high juice pH levels, wine instability, and poor red wine color. Whiting (2003) has reported K accumulation in ten commercial grape rootstocks. While Schwarzmann, 140 Ru, 99 R, and 101–14 Mgt showed moderate levels of K accumulation in the leaves, 420A, 110 R, 5C Teleki, 5BB Kober, 1103 P, and SO4, accumulated K at lower levels (Table 6.21). Boulay (1982) reported rootstocks that absorb the most K, 44–53 M and SO4, also absorb the least Mg.

Table 6.21 Accumulation of K in grape rootstocks (Whiting 2003)

Rootstocks that accumulate K at moderate levels	Rootstocks that accumulate K at lower levels
Schwarzmann	420A
140 Ru	110 R
99 Richter	5C Teleki
101–14 Mgt	5BB Kober
	1103 P
	SO4

When these rootstocks are used with Mg demanding scions, such as 'Grenache' or 'Cabernet Sauvignon', the result is a consequential Mg deficiency with its associated problems. Not only do recent studies into the influence of rootstock on mineral uptake contribute to future refinements of rootstock choices, but they are useful in interpreting past rootstock trials, especially where soil composition is known (Morton and Jackson 1988).

6.2 Cold Conditions

In general, grapevines grow well in temperate climates, which provide long, warm periods during the crucial flowering, fruit set, and ripening periods (Johnson and Robinson 2005). Grapes growing in cool climates are especially vulnerable to the negative effects of excessive vine vigor. Not only can the rootstock affect wine quality each year if ripening is delayed, but it can be a critical factor in winter survival. In cold climates, rootstock research and subsequent selection should be based on selecting for cold tolerance traits (Perry and Sabbatini 2015) including development of periderm lignification (early and good acclimation–cane maturation), the number of mature nodes during acclimation, leaf senescence and abscission (early and good acclimation), winter vine survival, winter injury on trunk and canes, and the number of injured buds (numbers of shootless nodes). Gender can be a specific concern as males show earlier acclimation given the commitment of female vines to crop ripening. In a study of V. riparia in southwestern Ontario, vines in the male population were larger than those in the female population (Rahemi et al. 2016), which may confirm that male vines are more tolerant to cold. Attributes in the spring that contribute to cold hardiness include late bud burst (late de-acclimation) and early veraison and precocity. Susceptibility to crown gall in spring, bud shape, and bud scale thickness, and ability to replace dead buds with latent buds after cold injury are other factors warranting consideration (Rahemi 2016). There appear to have been only a few studies related to rootstock effects on scion acclimation and cold hardiness.

Howell (2005) suggested short-cycle rootstocks would be most desirable for a cool climate. Perry and Sabbatini (2015) indicated that, while rootstocks do not have a direct effect on cold vine hardiness, short-cycle rootstocks can enhance cold acclimation of the vine. A short-cycle package with the potential of improving fruit quality at harvest is important in cool climate viticulture. Rootstocks that influence the scion with a short cycle and do not induce excessive vigor would help to avoid cold injury and ensure crops mature sufficiently in the fall. Howell (1987) reported that American varieties, as well as several French-American hybrids, were more productive and cold hardy following planting in grape replant sites when grafted onto the 3309 C rootstock and their cane hardiness is very good on this rootstock.

In the cool Finger Lakes region of New York, rootstocks with *V. berlandieri* parentage, such as 5BB and 5C, have tended to produce undesirable rank growth. In contrast, *V. riparia* × *V. rupestris* rootstocks, such as 3309 C and 101–14 Mgt, had less vegetative vigor and, therefore, has been used in this region more successfully (Pool 1987). In the temperate climate of Virginia, 'Cabernet Sauvignon' and 'Merlot' on 5BB are a poor combination due to excessive vegetative growth and subsequent winter injury, while White Riesling on 5BB is acceptable (Morton and Jackson 1988).

Some of the eastern Asian species, such as *V. amurensis*, are used increasingly in modern rootstock breeding programs for their cold tolerance (Alleweldt and Poss-ingham 1988), even though this diminishes the phylloxera resistance of the resulting crosses (Keller 2010). Rootstocks originating from *V. riparia* are often hardier to cold than other rootstocks, whereas rootstocks with *V. × champinii* as a parent are sometimes less cold-hardy (Stafne and Carroll 2006). Miller et al. (1988a) found differences in cane and bud hardiness of rootstock 5BB, 3309 C, and SO4. Root-stock 3309 C had the most cold-hardy canes and buds; its acclimation in fall was faster, and de-acclimation in spring was slower than that in 5BB and SO4. Miller et al. (1988b) also found that different rootstocks affected the cold hardiness of scions grafted to them. For example, grafted 'White Riesling' was significantly harder than own-rooted vines. Different rootstocks, however, had different effects on the cold hardiness of scion cultivars–the observed differences in LT50 values (the lowest temperature lethal to 50% of the primary buds) ranged from 0.5 to 3.0 °C in cane hardiness. 'White Riesling' grafted to 3309 C had significantly fewer shootless nodes, while on 5BB it had hardier buds. They proposed that 3309 C would be the most cold-hardy rootstock. Striegler and Howell (1991) compared grafted 'Seyval' on 5BB, 3309 C with own-rooted 'Seyval'. The rootstock did not affect the distribution of hardy canes within the canopy; 'Seyval' grafted on 3309 C appeared to be the hardiest. Palliotti et al. (1991) investigated the rootstock effects on frost sensitivity of grapevines 'Cabernet Sauvignon' and 'Chardonnay'. There was less damage to scions on 5BB and 1103 P than on SO4 and 420A. The development of replace-ment buds after cold damage was greater in 'Cabernet Sauvignon' on 1103 P than in 'Chardonnay' on 5BB (Gu 2010).

In the temperate climate and generally fertile soils of Virginia, Cabernet Sauvi-gnon and Merlot on 5 BB are a poor combination due to excessive vegetative growth and subsequent winter injury, while White Riesling on 5 BB, however, is acceptable (Morton and Jackson 1988).

References

Alleweldt G, Possingham JV (1988) Progress in grapevine breeding. Theor Appl Genet 75(5):669–673

Ambrosi H, Becker H (1978) Der Deutsche Wein. Gräfe & Unzer, München, p 91

Bavaresco L, Poni S (2003) Effect of calcareous soil on photosynthesis rate, mineral nutrition, and source-sink ratio of table grape. J Plant Nutr 26(10 & 11):2123–2135

Boulay H (1982) Absorption différenciée des cépages et des portegreffes en Languedoc. Prog. Agric. Vitic 99:431–434

Branas J, Vergnes A (1957) Morphologie du systeme radiculaire. Progres Agric Vitic. 1:3–13

Carbonneau A (1985) The early selection of grapevine rootstocks for resistance to drought conditions. Am J Enol Vitic 36:195–198

Champagnol F (1984) Eléments de physiologie de la vigne et de viticulture générale, Déhand, Montpellier, France

Christensen LP (2003) Rootstock selection. In: Christensen LP, Dokoozlian NK, Walker MA, Wolpert JA (eds.). ANR Pub 3419. Wine Grape Varieties in California. pp 12–15.

Cox C (2014) Rootstocks as a management strategy for adverse vineyard conditions, phylloxera and grape industry board of South Australia, Fact sheet 14. www.wineaustralia.com

de la Torre ID, del Campillo MC, Barrón V, Torrent J (2010) Predicting the occurrence of iron chlorosis in grapevine with tests based on soil iron forms. J Int Sci Vigne Vin 44:77–85

Drouineau G (1942) Dosage rapide du calcaire actif des sols. Nouvelles données sur la repartition et la nature des fractions calcaires. Annal Agron 12:441–450

Dry N (2005) Final report to grape and wine research and development corporation, phylloxera and grape industry board of South Australia, PGI 05/01, p 22

Dry N (2008) Grapevine rootstocks. Selection and management for South Australia vineyards. Phylloxera and grape industry board of South Australia. Lythrum Press, p 85

Dry N (2015) Pathway to successful rootstock use. https://mpva.com.au/exfiles/Rootstock_select ion.pdf (Mornington Peninsula Vignerons Association, Victoria, Australia)

Fregoni M (1977) Irrigation des vignobles. Problèmes de physiologie, debiochimie, de génétique, d'agronomie, de technologie et d'économie. Bul-letin de l'OIV551. pp 1–17

Galet P (1947) Résistance des porte-greffes à la chlorose. Progrès Agricole Et Viticole 36:128–133

Galet P (1998) Grape varieties and rootstock varieties. Oenoplurimedia press

Galet P (1998) Grape Varieties and Rootstock Varieties. Oenoplurimédia, Chaintré, France.

Gladstones JS (1992) Viticulture and environment: a study of the effects of environment on grape growing and wine qualities, with emphasis on present and future areas for growing wine grapes in Australia p 310

Gu S (2010) Effect of rootstocks on grapevines, Kentucky State University: http://www.pawpaw. kysu.edu/viticulture/Information/Rootstock

Guillon JM (1905) Étude générale de la vigne: historique les vignobles et les crus anatomie et physiologie, sol et climat. Masson, Paris, pp 191–194

Hamman RA, Savage JSD, Larsen H (1998) The Colorado grape growers' guide. Colorado State University, Colorado Agricultural Experiment Station: Fort Collins, Colorado. Bulletin 550A. p 85

Harmon FN, Snyder E (1934) Grape rootstock distribution study. Proc Amer Soc Hort Sci 32:370–373

Hofäcker W (2004) Ergebnisse und überlegungen zum einfluss der unterlage auf ertrag und qualität der rebe. deutsches weinbau-jahrbuch. Ulmer Verlag Stuttgart, pp 175–183

Howell GS (1987) Vitis rootstocks. In: Rom RC, Carlson RF (eds) Rootstocks for fruit crops. Wiley Inc., New York, pp 451–475

Howell GS (2005) Rootstock influence on scion performance. Grapevine rootstocks; current use, research and application. Proceedings of the 2005 rootstock symposium. Pub by MVEC, pp 47–55

Hunter JJ, Archer E, Van Schalkwyk D, Strever AE, Volschenk CG (2016) Grapevine roots: interaction with natural factors and agronomic practices. Acta Hortic 1136:63–80

Jackson RS (2000) Wine Science: Principles, Practice and Perception. 2nd ed. Academic Press, San Diego.

Johnson H, Robinson J (2005) The world atlas of wine. Mitchell Beazley, pp 20–21, ISBN 1-84000-332-4

Juste C, Pouget R (1972) Apprétiation du poivoir chlorosant des sols par un nouvel indice faisant intervener le calcaire actif et le ferrihydrite facilement extractible: application au choix des porte-greffes de la vigne. C R Acad Agric 58:352–357

Keller M (2010) Botany and anatomy, In: Keller M (ed). The science of grapevines, Anatomy and physiology. Academic Press, Elsevier Inc., San Diego

Lambert JJ, Anderson MA, Wolpert JA (2008) Vineyard nutrient needs vary with rootstocks and soils. Calif Agric 62(4):202–207

Lupascu N, Nicolaescu M, Chirila E (2009) Chlorosis risk evaluation from a new viticulture area located on a former Army region. In: Simeonov L, Hassanien MA (eds). Exposure and risk assessment of chemical pollution-contemporary methodology. Springer Media, pp 415–421

Miller DP, Howell GS, Striegler RK (1988a) Cane and bud hardiness of selected grapevine rootstocks. Am J Enol Vitic 39:55–59

Miller DP, Howell GS, Striegler RK (1988b) Cane and bud hardiness of own-rooted white riesling and scions of white riesling and chardonnay grafted to selected rootstocks. Am J Enol Vitic 39:60–66

Morton LT, Jackson LE (Jan 1988) Myth of the universal rootstock: the fads and facts of rootstock selection, Proceedings Second international cool climate viticulture and oenology symposium. Auckland, New Zealand

Ollat N, Peccoux A, Papura D, Esmenjaud D, Marguerit E, Tandonnet JP, Bordenave L, Cookson SJ, Barrieu F, Rossdeutsch L, Lecourt J, Lauvergeat V, Vivin P, Bert PF, Delrot S (2016) Rootstocks as a component of adaptation to environment. In: Gerós H, Chaves MM, Gil HM, Delrot S (eds) Grapevine in a changing environment: a molecular and ecophysiological perspective. https://doi.org/10.1002/9781118735985.ch4

Palliotti A, Cartechini A, Proietti P (1991) Influence of rootstock and height of training system on spring frost sensibility of chardonnay and cabernet Sauvignon grape cultivars in the Umbria region. Annali Della Facolta Di Agraria 45:283–291

Pavloušek P (2013) Tolerance to lime—induced chlorosis and drought in grapevine rootstocks. In: Vahdati K, Leslie C (eds) Abiotic stress: plant responses and applications in agriculture, Chap 9, pp 277–306, IntechOpen

Perold AI (1927) A treatise on viticulture. Macmillan, London

Perry RL, Sabbatini P (2015) Grape rootstocks for Michigan. Michigan State University, Department of Horticulture, vol E3298. pp 1–8. www.grapes.msu.edu

Pongrácz DP (1983) Rootstocks for Grape-vines. Cape Town, South Africa: David Philip Pub. p 150

Pool RM (July–Aug 1987) Rootstock and winter hardiness. Vineyard & Winery Management, p 40

Pouget R, Delas J (1989) Le choix des porte-greffes de la vigne pour une production de qualite. Connais. Vigne Vin, pp 27–31

Pouget R, Ottenwalter M (1978) Etude de l'adaptation de nouvelles variétés de porte-greffes à des sols très chlorosants. Connaissance De La Vigne Et Du 12:167–175

Progressive Viticulture (2011) Selecting a rootstock for a wine grape vineyard. http://www.progressivevit.com/selecting-a-rootstock-for-a-wine-grape-vineyard/. (Accessed Dec 2016)

Rahemi A (2016) Breeding of grapevine rootstocks for adaptability to the environmental stresses, third national conference of grape and raisin conducted by research institute of grape and raisin (RIGR), 27–29 September 2016. Malayer University, Malayer, Iran

Rahemi A, Dale A, Fisher H, Taghavi T, Bonnycastle A, Kelly J (2016) A report on Vitis riparia in Ontario, Canada. Acta Hortic 1136:33–38

Ravas L (1902) Les vignes Americaines: porte-greffes et producteurs directs. Coulet et Fils, Montpellier, p 376

Rusu M, Mihailescu T, Oroian I, Dumitras A, Marghitas M (2005) Agrochemistry (in Romanian). Ceres 2:370–389

Ryan J, Rashid A, Torrent J, Kwong Yau S, Ibrikci H, Sommer R, Erenoglu EB (2013) Micronutrient constraints to crop production in the Middle East-West Asia region: significance. Res, Manag, Adv Agron 122:1–84

Samson C, Castéran P (1971) Culture de la vigne. Sciences et tech-niques de la vigne. Tome 2—culture, pathologie, défense sanitaire de lavigne. Paris, France, Dunod

Schmid J, Manty F, Rühl EH (2015) Optimisation of sustainability of grapevine varieties by selecting rootstock varieties under different environmental conditions and creating new rootstock varieties. Retrieved 5 Oct 2019. http://www.viticulturasustentavel.com/pdfs/Prof_Joachim_Schmid.pdf

Scienza A, Boselli M (1981) Frequency and biometric characteristics of stomata in some grapevine rootstocks. Vitis 20:281–292

Serra I, Strever A, Myburgh PA, Deloire A (2014) Review: the interaction between rootstocks and cultivars (Vitis vinifera L.) to enhance drought tolerance in grapevine. Aust J Grape Wine Res 20(1):1–14

Shaffer RG (2002) The Effect of Rootstock on the Performance of the Vitis vinifera Cultivars Pinot Noir, Chardonnay, Pinot Gris, and Merlot. MS Thesis, Oregon State University, Corvallis

Shaffer R, Sampaio TL, Pinkerton J, Vasconcelos MC (2004) Grapevine rootstocks for Oregon vineyards. EM 8882, p 11. Oregon State University Extension Service, Corvallis, Oregon

Soar CJ, Loveys BR (2006) From sprinklers to drippers–impacts on grapevine root architecture. In: Walker RR (ed) Proceedings of a workshop on vine carbohydrate dynamics and source-sink relationships. CSIRO Plant Industry, Merbein, Australia, pp 66–75

Southey JM, Jooste JH (1991) The effect of grapevine rootstock on the performance of Vitis vinifera L. (cv. Colombard) on a relatively saline soil. S Afr J Enol Vitic 12(1):32–41

Southey JM (1992) Grapevine rootstock performance under diverse conditions in South Africa. Paper presented at: rootstock seminar: a Worldwide Perspective, Reno, Nevada, USA

Stafne ET, Carroll B (2006) Rootstocks for grape production. Oklahoma Cooperative Extension Service, HLA-6253, p 4

Striegler RK, Howell GS (1991) Influence of rootstock on the cold hardiness of Seyval grapevines. I. Primary and secondary effects on growth, canopy development, yield, fruit quality, and cold hardiness. Vitis 30:1–10

Striegler RK, Howell GS, Flore JA (1993) Influence of rootstock on the response of Seyval grapevines to flooding stress. Amer J Enol Viticult 44:313–319

Sweet N, Walker MA (2007) Early assessment of geotropic (gravitropic) root angle in grape root-stock as a method to predict vigor induction in the field, papers and posters presented at the ASEV 57th annual meeting 27–30 June 2006. Sacramento, California, Am J Enol Vitic 57(3):382–395

Taylor P (2002) Evaluation of Vitis Rootstocks for Tolerance to Low Soil pH. MS Thesis, Oregon State University, Corvallis

Tee E, Burrows D, Boland AM, Putland S (2003) Best irrigation management practices for viticulture in the Murray darling basin. Cooperative research centre for Viticulture, Adelaide

van Schalkwyk D (2017) Fruit production: get the right rootstock for planting grapes, Dec14, 2017. Retrieved: 9 Sep 2019. https://www.africanfarming.com/fruit-production-get-right-rootstock-pla nting-grapes

Whiting JR (2003) Selection of grapevine rootstocks and clones for Greater Victoria. Department of Primary Industries: Victoria, p 40

Whiting JR (2004) Rootstocks. In: Dry PR, Coombe BG (eds) Viticulture volume 1: resources. Winetitles, Adelaide

Winkler AJ, Cook JA, Kliewer WM, Lider LA (1974) General viticulture. University of California Press, Berkeley

Chapter 7
Rootstocks and Vineyard Design Considerations

Calculations of the quantity of vines to be ordered when establishing a new vineyard is dependent on fruiting variety, vine spacing, and training system (Morton 2007). Inter-row and intra-row spacing in a vineyard design depends on several factors, especially sunlight interception and enough space for adequate growth. Growth potential is related to vine vigor, and vigor refers to the rate of shoot growth or final shoot length. Vigor is sometimes considered as a vine size, not always directly related. Vine size includes leaves, shoots, trunks, and roots (Lakso 2013).

The growth variables predominantly considered in grape breeding are vigor, the formation of axillary shoots, and upright shoot growth. Regularly, vines with medium vigor and with a well-balanced ratio of vegetative and generative growth are preferred. Usually, poor or very weak growing seedlings become discarded in an early seedling stage. Since the vigor of grafted vines is also influenced to some extent by the rootstock, the vigor can be adjusted by combining a strong growing rootstock with a weak growing scion cultivar and vice versa. However, the evaluation of vigor should be determined in different locations with different rootstocks in an advanced breeding stage in order to identify interactions with soil type and rootstock variety (Eibach and Töpfer 2015).

Rootstocks clearly have a role in determining the vigor of their grafted scions. Yet, the relationship is poorly understood. This text confines the term 'vegetative vigor' to the growth of above-ground non-reproductive structures of the grapevine: shoots and wood. In doing so, vegetative vigor is separate from sexual reproductive vigor, growth of inflorescences and grape clusters, and also from root growth. An increase in vegetative vigor means that the amount of leaf and shoot material increases, and the number and size of both active and dormant meristems increases (Branas 1974). Associated with increased vigor is an increase in auxin, a plant hormone that prolongs the period of cellular expansion and allows the extension of the young parts of the shoot. Chlorophyll and chloroplasts in the leaves also increase. Vigorous shoots have more leaves, and they branch more strongly and profusely. It is as if one apex of the plant cannot use all the energy and resources available, so other apices join in the activity. Increased vegetative vigor often gives an increase of reproductive vigor; for

A. Rahemi et al., *Grape Rootstocks and Related Species*,
https://doi.org/10.1007/978-3-030-99407-5_7

Table 7.1 The growth rate of
5 years Zinfandel vines on
various rootstocks, St.
Helena, USA (Hayne 1896)

Rootstock	Growth		
	May	August	September
Rupestris 'Martin'	Vigorous	Good	–
Rupestris 'Mission'	Fair	Fair	Fair
Rupestris 'St. George'	Vigorous	Good	–
Riparia 'Gloire'	Strong	–	Fair
Riparia 'Grande Glabre'	–	–	Fair
Solonis	Weak	Fair	Fair
Vialla	Uneven	Weak	Fair
101–14 Mgt	Vigorous	Fair	Strong
3309 Mgt	Good	Fair	Strong

example, more floral buds are initiated (Champagnol 1984). But adverse phenological consequences can also result, e.g., the vegetative period of growth can be maintained for a longer time, veraison and maturity can be delayed. Excessive vigor also can have harmful effects on canopy shading and fruit ripening (Perry and Sabbatini 2015). When vegetative vigor is very high, serious problems arise. Rives (1971) describes some of these: "The consequences are a high incidence of post-bloom berry abortion or whole bunch abortion, an overall decrease in foliage efficiency due to mutual shading and to the blocking of photosynthesis from the premature closing of the stomata under the water stress caused by the very increase in leaf area, the build-up within the canopy of a micro-climate which is highly favorable to disease especially bunch rot (*Botrytis cinerea*) while the density of the canopy makes it harder for the sprays to reach their targets properly". Ideally, measurements of the rate of vegetative growth rather than the amount of vegetative growth would be the preferred way to evaluate vegetative vigor. With growth rates, changes in vigor at various parts of the season could be determined, e.g., between the pre-bloom, bloom, and fruit-maturation stages or between periods of high and low water availability. Rate measurements, however, demand large amounts of time, and they use a lot of destructively harvested material. Measurements of standing crops at the end of a growing season are more realistically achieved in the vineyard (Morton and Jackson 1988).

Fardossi et al. (1998) classified 31 rootstocks into five groups based on their vigor. Rupestris du Lot, 99 R and 3309 C were in the very low vigor group; Amos, 1103 P, 140 Ru, G1, G9, and 1616 C were in the low vigor group; Cosmo 10, 8B, 125AA, 420A, 725P, 41B, 333EM, Riparia Sirbu and 101–14 Mgt were in the middle vigor group; Fercal, Fercal 242, and Aripa (143A) were in the high vigor group; and 5BB, R27, 225 Ru, Riparia Portalis, G26, and Börner were in the very high vigor group.

Further studies have reported on the influence of rootstocks on grapevine vigor (Hayne 1896; Progressive Viticulture 2011; Southey 1992; Hunter et al. 2016; van Schalkwyk 2017) (Tables 7.1, 7.2, and 7.3).

Table 7.2 Rootstock influences on grapevine growth vigor (Progressive Viticulture 2011)

Very high	High	Moderate high	Moderate	Moderate low
Ramsey	125AA	SO4	1616 C	Schwarzmann
Dogridge	140 Ru	5C	RS-9	RGM
	Freedom	99 R		
	RS-3	110 R		
		1103 P		
		3309 C		
		101–14 Mgt		
		St. George		

(Copyright © 2011 Progressive Viticulture, http://www.progressivevit.com/selecting-a-rootstock-for-a-wine-grape-vineyard/)

Table 7.3 Vigor attributes of the most popular rootstocks in South Africa (Southey 1992; Hunter et al. 2016; van Schalkwyk 2017)

Rootstock	Vigority
99 R	Excellent
101–14 Mgt	Good
110 R	Excellent
140 Ru	Excellent
143-B Mgt	Excellent
420-A Mgt	Good
775 P	Excellent
1045 P	Good
1103 P	Excellent
3306 C	Good
3309 C	Good
Jacquez	Good
Ramsey	Excellent
Rupestris du Lot	Excellent
SO4	Good
USVIT 8–7	Excellent

References

Branas J (1974) Viticulture. Imprimerie Dèhan, Montpellier, p 990

Champagnol F (1984) Eléments de physiologie de la vigne et de viticulture générale. Déhand, Montpellier, France

Eibach R, Töpfer R (2015) Traditional grapevine breeding techniques. In: Reynolds A (ed) Grapevine breeding programs for the wine industry. Woodhead Publishing Series in Food Science, Technology and Nutrition: Number 268

Fardossi A, Stierschneider I, Schober V (1998) Uber die wuchskraft und das nahrstoffaneig-
nungsvermogen von 31 verschiedenen Unterlagsrebsorten in einer Schnittrebenanlage (Mutter-
garten). Mitteilungen Klosterneuburg 48:1–16

Hayne P (1896) Resistant vines, their selection, adaptation, and grafting, appendix to viticultural
report. University of California, College of Agriculture, Agricultural experiment station

Hunter JJ, Archer E, Van Schalkwyk D, Strever AE, Volschenk CG (2016) Grapevine roots:
interaction with natural factors and agronomic practices. Acta Hortic 1136:63–80

Lakso A (2013) Untangling the concepts of vine size, capacity, crop level, vigor, and vine balance.
Cornell CALS. Newsletter, Issue 13, March 2013. Retrieved 20 July, 2021. https://grapesand
wine.cals.cornell.edu/newsletters/appellation-cornell/2013-newsletters/issue-13/grapes-101/

Morton LT, Jackson LE (Jan 1988) Myth of the universal rootstock: the fads and facts of rootstock
selection. In: Proceedings second international cool climate viticulture and oenology symposium.
Auckland, New Zealand

Morton LT (2007) Rootstock review, how to set priorities when choosing rootstocks. Wines and
Vines

Perry RL, Sabbatini P (2015) Grape rootstocks for Michigan. Michigan State University, Department
of Horticulture, E3298:1–8. www.grapes.msu.edu

Progressive Viticulture (2011) Selecting a rootstock for a wine grape vineyard. http://www.progre
ssivevit.com/selecting-a-rootstock-for-a-wine-grape-vineyard/. (Accessed Dec 2016)

Rives M (1971) Statistical analysis of rootstock experiments as providing a definition 2 of the terms
vigour and affinity in grapes. Vitis 9:280–290

Southey JM (1992) Grapevine rootstock performance under diverse conditions in South Africa.
Paper presented at: rootstock Seminar: a worldwide perspective. Reno, Nevada, USA

van Schalkwyk D (2017) Fruit production: get the right rootstock for planting grapes, 14 Dec 2017,
Retrieved: 9 Sep 2019. https://www.africanfarming.com/fruit-production-get-right-rootstock-pla
nting-grapes

Chapter 8
Function of Grape Rootstock

8.1 Effect of Rootstock on Fruit Set and Yield

Crop load is the balance of fruit mass and canopy size. It is typically calculated using yield at harvest and comparing it to the dormant pruning weight measured in the winter following harvest (i.e., Crop load = Vine yield/dormant pruning weight). This calculation is known as the Ravaz Index (Skinkis and Vance 2013) (Tables 8.1, 8.2, and 8.3).

8.2 Effect of Rootstock on Berry Size

Rootstocks can affect berry size, but the results vary by location. For example, berry weight (g/100 berries) of 'Pinot noir' on different rootstocks was evaluated at two locations by Schmid et al. (2015) and was ranked (highest to lowest) for the rootstocks Teleki 8B, Börner, 110 R, 3309 C, 161-49C, Sori and 101–14 Mgt at one location, but as Börner, Teleki 8B, 110 R, 101–14 Mgt, Sori and 3309 C at the second location. Berry growth in grapevine occurs in a double sigmoid curve pattern at three phases. In phase I (berry formation), berry size increases rapidly through both cell elongation and an increase in cell number, and organic acids, tannins, minerals, and other substances rapidly accumulate in the fruit. During phase II or lag phase, cell expansion does not occur, and organic acids peak before being degraded or diluted acid depending. In phase III, berry ripening continues and berry expansion resumes through cell elongation and soluble solid accumulation. Also, during phase III, anthocyanins accumulate in the berry skins, glucose, and fructose content increases, and organic acid concentration continues to decline (Kennedy, 2002). Insufficient water reduces berry size (Iland et al. 2011), then it is important to choose proper drought-tolerant rootstocks for dryland areas (Rahemi 2016).

Table 8.1 Classification of grape rootstocks based on their effects on fruit set (Dry 2015)

Classification	Rootstock
Improves fruit set and or fertility	5C Teleki
	Schwarzmann
	101–14 Mgt
	SO4
	3309 C
	420A
	Riparia Gloire
Reduces fruit set*	1103 P
	110 R

*not recommended for poor set varieties

Table 8.2 Crop rate of 5 years Zinfandel vines on various rootstocks, St. Helena, US (Hayne 1896)

Rootstock	Crop
Rupestris 'Martin'	Large
Rupestris 'Mission'	Good
Rupestris 'St. George'	Good
Riparia 'Gloire'	Good
Riparia 'Grande Glabre'	Fair
Solonis	Large
Vialla	Fair
101–14 Mgt	Good-heavy
3309 C	Heavy

Table 8.3 Comparison of crops on *Vitis riparia* and *V. rupestris* (Hayne 1896)

Cultivar	Crop of vines on	
	V. riparia	*V. rupestris*
Kleinberger	Poor	None
Sultana	Good	Fair
Chardonnay	Good	Fair
Valdepenas	Very large	Poor
Gros Mansenc	Heavy	Fair
Fresa	Fair	Fair
Marsanne	Fair to large	Fair
Vernaccia	Fair to large	Fair
Cornichon	Fair	Good

References

Dry N (2015) Pathway to successful rootstock use, https://mpva.com.au/exfiles/Rootstock_select
ion.pdf (Mornington Peninsula Vignerons Association, Victoria, Australia)

Hayne P (1896) Resistant vines: their selection, adaptation, and grafting. Appendix to viticultural
report, University of California, College of agriculture, Agricultural experiment station

Iland PG, Dry P, Proffitt T, Tyerman S (2011) The Grapevine: from science to the practice of growing
vines for wine. Patrick Iland Wine Promotions Pty Ltd, Adelaide

Kennedy JA (2002) Understanding grape berry development. Practical winery and vineyard
management, 14–19

Rahemi A (2016) Breeding of grapevine rootstocks for adaptability to the environmental stresses,
third national conference of grape and raisin conducted by research institute of grape and raisin
(RIGR), 27–29. Malayer University, Malayer, Iran

Schmid J, Manty F, Rühl EH (2015) Optimisation of sustainability of grapevine varieties by selecting
rootstock varieties under different environmental conditions and creating new rootstock varieties.
http://www.viticulturasustentavel.com/pdfs/Prof_Joachim_Schmid.pdf Accessed 5 Oct 2019

Skinkis PA, Vance AJ (2013) Understanding vine balance: an important concept in vineyard
management, Oregon state university, Extension catalog https://catalog.extension.oregonstate.
edu/em9068

Chapter 9
Choosing Grape Rootstock

Growers must be careful when ordering new rootstock-scion combinations and be willing to pay more for rare or difficult-to-propagate rootstocks. Nurseries faced with new rootstocks need to study ways to increase their percentage of successful grafts; chances are the "new" rootstock for them is common in another country. Also, there is a tendency to favor rootstocks that produce strong early growth over those that take longer to get established. This can be a mistake because often, the "slow starters" are the best producers in the long term (Morton and Jackson, 1988). It is important to know about the planting site characteristics, including soil characteristics (soil series, texture, depth to hardpan or clay layer, percent organic matter, soil pH, irrigation water quality and salinity, soil water holding capacity, water drainage, soil nutrient status), and topography. Also, one must know the objectives such as scion cultivar, market, solid-set, or replant sites and then can choose the rootstock based on the information (Castle and Ferguson 2003).

Rootstock selection should start with agronomic considerations such as survival (phylloxera, nematode, cold tolerance) and adaptability (to soil structure and chemistry, water availability, vigor potential) (Morton 2007). Figure 9.1 includes rootstocks traditionally used in California as determined by the Progressive Viticulture website (2011). This chart was established with a focus on soil-borne pest presence, followed by growth factors, and environmental considerations should then be taken into consideration.

In further refinement, a software has been developed by Wine Australia (2021) to assist in grapevine rootstock selection. This software guides users through a series of multiple-choice options/questions that require the characteristics of a site to be defined and to better understand vine management requirements. The software asks for: How would you describe your region's climate (cool climate, warm climate, and hot climate), and how much irrigation water you are likely to have (ample irrigation, limited irrigation, and no irrigation)? What is the desired level of vine vigour (low, moderate, and high)? What is the salt content of your soil (non saline, slightly

A. Rahemi et al., *Grape Rootstocks and Related Species*,
https://doi.org/10.1007/978-3-030-99407-5_9

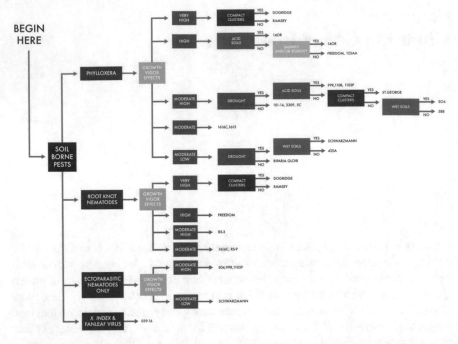

Fig. 9.1 1-The rootstock selection chart includes only rootstocks traditionally used in California. There are other rootstocks that have been used elsewhere in the world and are available in California that have yet to be tested. These include 775P, 779P, 5A, 8B, 161-49C, 225R, 41B, 1045P, 171–6, and K51-32; 2-This chart represents the best information and experience available at the time of its compilation; 3- assumes all plant materials are CDFA certified; 4- Copyright © 2009 progressive viticulture, with copyright permission received from Stan Grant, provit@earthlink.net; Ref: progressive viticulture 2011, http://www.progressivevit.com/selecting-a-rootstock-for-a-wine-grape-vineyard/

saline, moderately saline, and very saline)? How does the soil drain at your vineyard site (poorly, or well)? What is the pH of your soil (strongly acidotic, weakly acidotic, neutral, weakly alkaline, and strongly alkaline)? Do you need rootknot nematode tolerant rootstocks for your soil type (yes or no)? The software then uses the information to identify rootstocks for consideration.

Nursery guides are also commonly utilized to assist with rootstock selection (Table 9.1). These nursery guides provide basic, somewhat anecdotal guidance of the characteristics that are attributed to grape species and commonly used rootstocks as they relate to the tolerance of abiotic and biotic stressors in a particular location.

Table 9.1 Influences of commercial grapevine rootstocks on characteristics of scions and their tolerance to abiotic and biotic stresses

Rootstock	On scion characteristics		Vegetative cycle	On tolerance to abiotic stresses					
	Vigorous	Ease of propagation		Cold tolerance	Drought tolerance	Wet soil tolerance (Bad drainage)	Salinity tolerance (Salt)	Alkalinity tolerance (Lime)	Acid soil tolerance
O39-16	H	VL			L	M	L	L	
101–14 Mgt	M	H	AD	H	L	L	M	L	M
1103 P	H	H	DE	L-M	M-H	M	M-H	M-H	L-M
110 R	M-H	L-M	DE		M-H	L	M	M	M
125AA	H							H	
1202C	H								
140 Ru	H	M	DE	M	H	L	H	H	H
143B	M-H		MA		M-H	H	M-H		
161–49 C	M-L		AD	M	M-L	L		H	
1616 C	M	H	AD		L	M	M	L	
1616 E	L								
18–815 Cl	M								
196–17 Cl	H				H	L		L	
3309 C	L-M	H	AD	H	M	L-M	L	L	
333EM	H		N		M	H		H	
34EM	H		AD		H	H		M-H	
41B	M			M	M	L		H	
420A	L-M	M	AD	H	L	M	L	M-H	
44–53 M	M	H	AD		M	M	L	L	M
5BB	M	H	DE	L	M	L		M-H	
5C	L-M	H		L	L	L-M	M	M-H	
775 P				M					

(continued)

Table 9.1 (continued)

Rootstock	On scion characteristics			On tolerance to abiotic stresses					
	Vigorous	Ease of propagation	Vegetative cycle	Cold tolerance	Drought tolerance	Wet soil tolerance (Bad drainage)	Salinity tolerance (Salt)	Alkalinity tolerance (Lime)	Acid soil tolerance
99 R	M-H	M	DE		M-H	L	M	M	
AXR1	H								
Binova	H		N		M-H	H		M-H	
Börner	H				H	M		M	
Dog Ridge	VH	L			H	L-M	M-H	M	
Fercal	M		N		L-M	H		VH	
Freedom	H	M-H	DE		H	L	M	M	
Gravesac	M		AD		M	H		M	
GRN-1	M-H				M		M-H		
GRN-2	M				M		M		
GRN-3	M-H				M-H		M		
GRN-4	M-H				H		M		
GRN-5	H				H		M		
Harmony	H	H			M-H	L	L-M	M	
Ramsey	H	L	DE	M	H	L-M	M-H	M-H	
RGM	L	H	AD	VH	L	M	L	L	
RS -9	L								
RS -3	M-H								
Schwarzmann	M	H	AD		M	M	M-H	M	
SO4	L-M	M	MA	M	L-M	M	L-M	M-H	M
St. George	H	H			VA	L-M	M-H	L	
Teleki 8B	M				M			M	
US 8-7	M-H		MA		M-H	H	M-H		

(continued)

Table 9.1 (continued)

Rootstock	On tolerance to biotic stresses						
	Phylloxera tolerance	Nematode tolerance	Root-knot nematode	Dagger nematode	Root lesion nematode	Crown gall	Phytophthora cinnamomi
O39-16	H		L	H			
101–14 Mgt	M	L-M	M-H	M	M	H	L-M
1103 P	H	L	M	L		L	VL-M
110 R	H	L	L	L	L	L	L-H
125AA							
1202C							
140 Ru	H	L	L	L		L	L
143B			L-M	L-M	L-M		H
161–49 C							
1616 C	H	L	M-H	M-H			L
1616 E							
18–815 Cl							
196–17 Cl							
3309 C	H	L-M	L	L		H	L
333EM							
34EM							
41B							
420A	H		L-M	L			
44–53 M	H	M	L				
5BB	H	M	M-H	M			
5C	H		M	M			
775 P	H						
99 R	H		M	L-M	L-M		VL

(continued)

Table 9.1 (continued)

Rootstock	On tolerance to biotic stresses						
	Phylloxera tolerance	Nematode tolerance	Root-knot nematode	Dagger nematode	Root lesion nematode	Crown gall	Phytophthora cinnamomi
AXR1	L						
Binova							
Börner							
Dog Ridge	M-H		H	M			
Fercal							
Freedom	L	H	H	H			
Gravesac	H		L				
GRN-1	VH		VH	VH			
GRN-2	VH		VH	VH			
GRN-3	H		VH	VH			
GRN-4	H		VH	VH			
GRN-5	M-H		VH	VH			
Harmony	L		L-M	M-H			
Ramsey	H		H	M	VH		VH
RGM	H	M	M-H	M			
RS -9			H	H			
RS -3			H	H			
Schwarzmann	H	M-H	M	M-H			
SO4	H	L	M	M		L	L–H
St. George	H	VA	L	L			
Teleki 8B	H		H				
US 8–7			H				H

H = high, good; M = moderate, medium, average; V = very; L = low, poor, weak; S = semi; VH = very high, excellent; VL = very low; AD = advancing; MA = moderate advancing; N = no effect, VA = variable (Adapted and derived from Hedrick 1908; Wagner 1945; Pongrácz 1983; Main et al. 2002; Christensen 2003; Stafne and Carroll 2006; Dry 2007; Keller 2010; Peccoux 2011; Harris 2013; Teubes 2014; Walker et al. 2014; Perry and Sabbatini 2015; Wine Growers 2019)

References

Castle WS, Ferguson JJ (2003) Considerations for choosing the right rootstocks. University of Florida, IFAS Extension, HS932, http://edis.ifas.ufl.edu

Christensen LP (2003) Rootstock selection. In: Bettiga LJ (ed) Wine grape varieties in California. University of California, Oakland, CA, USA, pp 12–15

Dry N (2007) Grapevine rootstocks: selection and management for South Australian vineyards. (Lythrum press in association with phylloxera and grape industry board of South Australia: Adelaide)

Harris JL (2013) Effect of rootstock on vegetative growth, yield, and fruit composition of Norton grapevines. MSc Thesis, University of Missouri, p 95

Hedrick UP, Booth NO, Taylor OM, Wellington R, Dorsey MJ (1908) The grapes of New York. New York (State). Dept. of Agriculture—Agricultural Experiment Station. Albany: J.B. Lyon Co., state printers, p 788

Keller M (2010) Botany and anatomy, In: Keller M (ed). The science of grapevines, Anatomy and physiology. Academic Press, Elsevier Inc., San Diego

Main G, Morris J, Striegler K (2002) Rootstock effects on chardonel productivity fruit and wine composition. Am J Enol Vitic 53:37–40

Morton LT (2007) Rootstock review. How to set priorities when choosing rootstocks, wines and vines

Morton LT, Jackson LE (1988) Myth of the universal rootstock: the fads and facts of rootstock selection. In: Proceedings Second international cool climate viticulture and oenology symposium, Auckland, New Zealand

Peccoux A (2011) Molecular and physiological characterization of grapevine rootstock adaptation to drought. University of Bordeaux, PhD. dissertation, p 184

Perry RL, Sabbatini P (2015) Grape rootstocks for Michigan. Michigan State University, Department of Horticulture, E3298:1–8, (www.grapes.msu.edu)

Pongrácz DP (1983) Rootstocks for grapevines. David Philip, Cape Town, South Africa

Progressive Viticulture (2011) Selecting a rootstock for a wine grape vineyard. http://www.progre ssivevit.com/selecting-a-rootstock-for-a-wine-grape-vineyard/. Accessed Dec 2016

Stafne ET, Carroll B (2006) Rootstocks for grape production. Oklahoma cooperative extension service, HLA-6253, 4 p

Teubes A (2014) History of rootstocks in South Africa. Practical in the vineyard, Wineland Media, http://www.wineland.co.za/wp-content/uploads/2017/01/andrew4.jpg. Accessed Sep 14 2019

Wagner PM (1945) A wine-grower's Guide. Alfred A. Knopf, New York

Walker R, Blackmore D, Clingeleffer P, Walker MA, Gilliham M, Henderson S, Francis L, Holt H, Pearson (2014) Delivering chloride and sodium excluding rootstocks for quality wine production. Grape and wine research and development corporation (wine Australia), CSIRO, final report project number CSP1002

Wine Australia (2021) Grapevine rootstock selector tool. http://www.grapevinerootstock.com/

Wine Growers (2019) Rootstocks for grafted vines. http://www.winegrowers.info/rootstocks/home. htm

Chapter 10
Nursery Stock Certification and Testing Programs

Classical grapevine grafting methods on a commercial scale were started around 130 years ago (Grohs et al. 2017). Several certification schemes have been developed regionally to address the key issues of pathogen testing and variety/clone identification, but there is no unified global standard. EPPO (European and Mediterranean Plant Protection Organization) certification for the European Union is based on checking the presence of viruses in rootstocks. The APFIP (Australian Pome Fruit Improvement Program) in Australia, P PECP (Plant Protection Export Certification Program) in Canada, Entav-Inra (The Etablissement National Technique pour l'Amelioration de la Viticulture—Institut National de la Recherche Agronomique) in France, and FPS (Foundation Plant Services) in the United States program with verification protocols and technical recommendations from the production of nuclear material to commercial nurseries (Grohs et al. 2017).

10.1 Pathogen Testing and Certified Grapevine Rootstock

Grapevine species and hybrids used as rootstocks are hosts of about 65 viruses, eight viroids, and 13 phytoplasmas (Martelli 2014). In addition, some insects and nematodes are efficient natural vectors of certain viruses and phytoplasmas, making the situation even more complex (Fiore 2015). Recently, some new viruses have been discovered in grapevines and have rapidly become known as "emerging viruses" (Martelli and Saldarelli 2015). Such new diseases have been found in mother plants and, in a number of instances, have become widely disseminated among producing regions from propagation materials (Gramaje and Marco 2015). The criteria of phytosanitary quality in mother plants and cuttings need revision due to the increase in the incidence of fungi causing trunk diseases that cause the decline and death of young vineyards (Stamp 2001). In addition to viruses and fungi, bacterial diseases are another phytopathogenic group that can infect host plants and be transmitted via propagative material (Rodrigues Neto et al. 2011). Diagnosis is a fundamental step,

© The Author(s), under exclusive license to Springer Nature Switzerland AG 2022   101
A. Rahemi et al., *Grape Rootstocks and Related Species*,
https://doi.org/10.1007/978-3-030-99407-5_10

Table 10.1 Main indicators for virus and virus-like diseases of grapevine (EPPO 2008)

Indicator	Disease identified
Vitis rupestris St. George	Degeneration, fleck, *rupestris* stem pitting
Vitis vinifera Cabernet franc, Pinot noir, and other red cultivars	Leafroll
Kober 5BB (*Vitis berlandieri* × *Vitis riparia*)	Kober stem grooving
LN 33 (1613 C × *Vitis berlandieri*)	Corky bark, enation, LN33 stem grooving
Vitis riparia Gloire de Montpellier	Vein mosaic
110 R (*Vitis rupestris* × *V. berlandieri*)	Vein necrosis

EPPO (2008) provides full details of the conditions for the tests and suggests some alternative indicators

together with epidemiological studies for the implementation of effective control of diseases caused by viruses, viroids, phytoplasmas, bacteria, and fungi related to trunk diseases. Field observation is the first step in the phytosanitary checking process (Fiore 2015). Diagnosis of viruses, viroids, and phytoplasmas, indexing by mechanical transmission of the pathogen to herbaceous hosts, or by graft indexing in woody indicators, are established. For virus testing, visual inspection and indexing (testing on *Vitis* indicators, Table 10.1) is a simple method, but such indexing is not exclusive because there are some viral diseases (leafroll and rugose complex) that cannot be identified except on woody hosts. Some viruses which need a graft transmission test using specific indicators are: fanleaf (GFLV); grapevine degeneration (European nepoviruses); leafroll (GLRaV's), including GLRaV 1, 2, 3; Rugose wood complex (GVA, GVB); and Fleck (GFkV)1 (EPPO 2008).

Serology-based diagnosis, as an enzyme-linked immunosorbent assay (ELISA), is now the most commonly used test for viruses in grape rootstocks (Osman et al. 2013). ELISA tests have also been used as a screening process (Lázár 2003) (Fig. 10.1).

The use of ELISA is recommended for diagnosis of grapevine fanleaf virus and other European nepoviruses where they occur, Closteroviridae (GLRaV 1–9), vitiviruses (GVA and GVB), and maculavirus (GFkV), for which antisera are available. ELISA is reliable for identification of viruses belonging to the family Closteroviridae in American rootstocks, especially *Vitis rupestris* and its hybrids (EPPO 2008). Molecular diagnosis, using Next Generation Sequencing (NGS), allows the identification of all viruses, viroids, and phytoplasmas (Al Rwahnih et al. 2015). Conventional PCR, nested-PCR, and real time-PCR are being developed to identify and detect the causal agents of some diseases, such as Petri disease (Alaniz et al. 2009). These molecular techniques can also help the detection of new pathogens (Martelli 2014). RT-PCR probes have been recommended for virus testing (for GFLV,

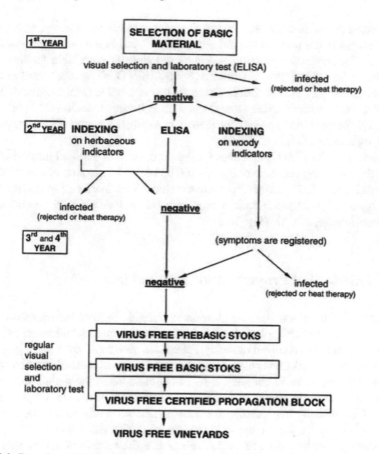

Fig. 10.1 Procedure of virological screening of grapevine in Hungary (Lázár, 2003)

GLRaV 1, 2, and 3, GVA, GVB, and GFkV) (EPPO 2008). PCR has gained popularity due to its higher sensitivity over bioassay and ELISA. Using RT-PCR in the single or multiplex reaction is strongly suggested on ELISA negative samples and can increase the number of positive results, especially for GFLV and GFkV detection.

These tests are widely used in practice. For example, all clones of rootstocks at Geisenheim University are routinely tested by serological testing methods (ELISA) for GFLV, ArMV, GLRaV1, GLRaV2, GLRaV3, GVA, and GFkV viruses, and are also tested using PCR for LR1, LR2, LR2 Red Globe (Grapevine rootstock stem lesion associated virus), LR3, LR4, LR5, LR9, GVA, GVB, GFkV (variants A and B), GFLV, and RRSV (Schmid et al. 2015).

Management of new diseases needs procedures to control viruses, viroids, and phytoplasmas in the process of cutting production and based on the prevention of infection. For sanitation procedures, some guidelines are available for heat treatment, and Shoot (meristem) tip culture in vitro (EPPO 2008) and clonal cleaning by thermotherapy and/or meristem culture are recommended (Maliogka et al. 2009). Control of vectors (nematodes, mealybugs such as Pseudococcidae, etc.) and weeds (hosts of pathogens and vectors) contribute to reducing the spread of viruses, viroids, and phytoplasmas (Tsai et al. 2010).

Based on EPPO (2008) guidelines for the production of certified grapevine rootstocks, the following successive steps should be taken: 1. selection of candidates for use as nuclear stock, 2. production of nuclear stock, 3. maintenance of nuclear stock, 4. production of propagation stock, 5. production of certified plants, 6. administration of the certification scheme (Fig. 10.2).

10.2   Intellectual Property and Plant Patents

In recent years there has been an increase worldwide in patented grape rootstock selections. Since the technology involved in both grapevine identification and plant disease detection has changed radically, there have been rapid changes in grapevine quarantine, clean stock management, and the certification of both scion and rootstock material. One of the most complex areas of negotiations in the global marketplace of grape rootstocks is related to intellectual property as these can be proprietary, protected by patents, trademarks, or contracts. Patented rootstocks might incur a charge or royalty fee for the use of that material during the lifetime of the patent. Some patents can be managed exclusively. After the release of a new rootstock, the speed with which it can be distributed nationally and internationally depends on the management of property rights, and to phytosanitary and quarantine laws of the importing country or region. In some countries, new varieties must also be extensively tested before they can be authorized for planting (Golino 2000).

10.3   Genetic Marker Testing for Rootstock Identification

Rootstock variety and clone identification is an important component of certification. Many different commercial rootstocks can be derived from the same parentage and can have a similar appearance. Consequently, identification based on ampelography (visual characteristics of grapevines, often only leaves) may not be suitable for closely related rootstocks. Molecular markers have been used to distinguish genetically distinct but phenotypically similar species, hybrids, and rootstocks. These methods have included: restriction fragment length polymorphism-RFLP (Bourquin et al. 1991, 1992, 1995), isoenzymes (Boursiquot and Parra 1992; Walker and Liu 1995), random amplified polymorphic DNA-RAPD (This et al. 1997; Bisztray et al.

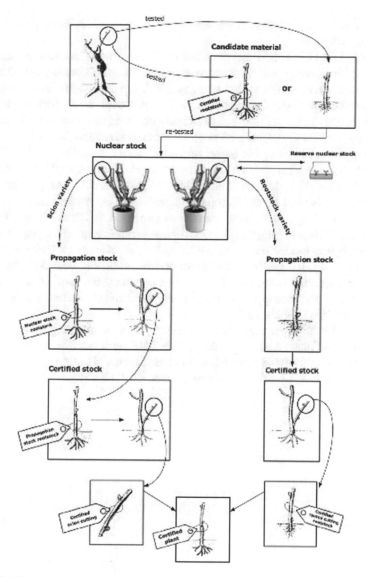

Fig. 10.2 Diagram of the stages in the grapevine certification scheme (EPPO 2008)

2003; Halasz et al. 2004), simple sequence repeats-SSR (Lin and Walker 1997, 1998; Sefc et al. 1999; Di Gaspero et al. 2000; Lowe and Walker, 2006; This et al. 2006; Moravcova et al. 2006; de Andrés et al. 2007; Upadhyay et al. 2007; Laucou et al. 2008; Crespan et al. 2009; Leão et al. 2009; Jahnke et al. 2011), amplified fragment length polymorphism-AFLP (de Andrés et al. 2007; Upadhyay et al. 2007), expressed sequence tag-simple sequence repeats, EST-SSR (Fan et al. 2014), and single nucleotide polymorphism-SNPs (Miller et al. 2013; Bianchi et al. 2020; D'onofrio 2020).

SSR markers (also called microsatellite markers) provide an excellent means of rootstock identification, new variety fingerprinting and assist in the maintenance of grape germplasm collections (Thomas et al. 1994; Lin and Walker 1998; Sefc et al. 1999; Dangl et al. 2001; Crespan et al. 2009). Although many such markers are available, a common set of nine pairs of highly polymorphic microsatellite markers are usually used for the identification of rootstocks (Maul and Töpfer 2015; VIVC 2019). These are di-nucleotide repeat sequence motifs, including VVS2 (Thomas and Scott 1993), VVMD5, VVMD7 (Bowers et al. 1996), VVMD25, VVMD27, VVMD28, VVMD32 (Bowers et al. 1996), VrZAG62 and VrZAG79 (Sefc et al. 1999). This set is used by the European GrapeGen06 consortium as the standard set for the screening of grapevine collections (Bacilieri and This 2010). The Vitis International Variety Catalogue (VIVC 2019), an online database for grape varieties maintained by the Julius Kuhn-Institut Federal Research Centre for Cultivated Plants (JKI) in Germany, has collected the microsatellite profiles of over 5000 grapevine varieties in a searchable database for easy comparison (Microsatellite data of varieties). A list of common grape rootstocks with their microsatellites characteristics has been provided in Table 10.2.

SNP profiles for more than 1600 grapevine varieties including rootstocks also available in VIVC database (VIVC 2021). Their SNP profiles extracted from different sources, including (1) the 48 plus 24 SNP markers developed by Cabezas et al. (2011), and (2) 14 plus 26 SNP loci proposed by Laucou et al. (2018).

Table 10.2 Common grape rootstock relationships based on nine microsatellites (Retrieved form Moravcova et al. 2006; VIVC 2019)

Rootstock	Clones	VVS2		VVMD5		VVMD7		VVMD25	
		A1	A2	A1	A2	A1	A2	A1	A2
Amos*		132	142	238	248	238	262		
Berlandieri colombard 2		141	143	223	234	231	249	252	254
Berlandieri Resseguier 2		147	149	228	238	231	233	247	249
Binova		145	147	238	268	233	265	239	249
Börner		131	151	232	268	231	245	235	251
Castel	196–17	133	145	268	268	249	264	239	241
	216–3	137	141	238	270	251	261	237	239
	4010	137	161	236	266	243	251	237	237
Cina		143	147	240	268	231	251	249	255
Cosmo	2	145	149	238	268	233	251		
	10	145	147	238	268	233	265	239	250
Colombard	2	143	151	234	242	239	239	249	255
Couderc	1613	129	139	240	270	243	251		
	1616	139	141	266	270	239	251	239	243
	3306	133	159	248	266	245	259	237	239
	3309	123	161	254	266	245	259	237	239
	161–49	137	141	230	268	231	251	237	269
Crăciunel*	2	138	142	250	250	248	258		
Dog Ridge		133	141	266	266	233	251	235	247
EM	34	135	143	242	266	231	251	249	267
	333	137	147	217	229	231	239	238	250

(continued)

Table 10.2 (continued)

Rootstock	Clones	VVS2		VVMD5		VVMD7		VVMD25	
		A1	A2	A1	A2	A1	A2	A1	A2
Fercal		143	143	238	266	231	251	237	255
Freedom		129	133	266	270	243	251		
Golia		137	145	266	270	249	257	237	241
Gravesac		141	161	266	268	245	251	237	239
Grézot	1	137	151	254	270	239	249	237	263
Harmony		133	139	266	270	243	251		
Kober	5BB	141	149	238	268	233	265	237	247
	125AA	141	147	228	268	231	251	237	249
Kraus1*		134	134	226	250	252	262		
LN 33		129	145	236	240	251	253	239	249
Malégue	44–53	139	145	254	266	233	239	239	243
Mgt	101–14	133	143	258	268	243	251	237	239
	420A	127	137	240	266	231	263	239	271
	41B	135	143	228	230	231	239	255	255
NAB		137	151	242	248	231	245	237	239
NOAH		125	129	252	252	235	255		
Paulsen	775	137	147	228	238	231	261	237	249
	779	137	149	238	270	231	257	237	247
	1045	137	149	228	238	233	257	237	247
	1103	137	147	238	238	233	257	237	249

(continued)

Table 10.2 (continued)

Rootstock	Clones	VVS2		VVMD5		VVMD7		VVMD25	
		A1	A2	A1	A2	A1	A2	A1	A2
Paulsen	1447	137	147	228	270	233	261	237	249
Ramsey		137	139	264	264	233	261	239	263
Richter	31	139	143	260	266	251	251	237	255
	57	137	143	248	270	231	257		
	99	137	149	238	238	231	261	237	247
	110	137	143	236	270	231	257	237	263
Rici		131	151	232	268	231	245	235	251
RGM		141	145	268	268	251	265	237	239
R.S.B	1	139	147	223	263	233	264	236	246
Ruggeri	140	137	143	248	270	231	257	237	263
Rupestris du lot		137	137	238	270	257	261	237	237
Schwarzmann		137	145	254	268	251	265	239	253
SO4		143	145	234	263	233	264	238	249
Sori		143	155	268	270	251	251	239	255
Sorisil		151	159	228	254	243	251	235	249
Teleki	8B	141	149	228	268	233	251	237	247
	5C	145	149	238	268	231	265	239	247
Vialla		125	135	268	268	235	251	235	235

(continued)

Table 10.2 (continued)

Rootstock	Clones	VVMD27		VVMD28		VVMD32		VRZAG62		VRZAG79	
		A1	A2	A1	A2	A1	A2	A1	A2	A1	A2
Amos*		178	204					187	199	251	255
Berlandieri colombard 2		252	254	233	247	243	249	192	194	250	252
Berlandieri Resseguier 2		192	204	236	252	260	260	210	214	251	251
Binova		204	212	216	236	260	260	200	214	251	255
Börner		186	208	216	284	240	240	174	190	255	255
Castel	196–17	190	208	216	244	240	240	188	200	255	261
	216–3	208	212	220	242	235	244	190	196	255	263
	4010	208	208	236	244	250	250	188	190	255	255
Cina		186	204	216	260	240	240	192	204	251	255
Cosmo	2	192	208	244	252	234	260	192	214	251	255
	10	192	212					200	214	251	259
Colombard	2	176	182	246	248	256	268	188	196	243	247
Couderc	1613	178	210	234	244			188	190	257	267
	1616	196	212	242	250	244	244	190	191	257	257
	3306	187	216	236	246	250	272	180	190	255	255
	3309	186	212	240	246	236	236	180	190	255	257

(continued)

Table 10.2 (continued)

Rootstock	Clones	VVMD27		VVMD28		VVMD32		VRZAG62		VRZAG79	
		A1	A2	A1	A2	A1	A2	A1	A2	A1	A2
Couderc	161–49	188	212	216	242	266	266	192	196	255	259
Crăciunel*	2	204	204					201	201	253	253
Dog Ridge		192	214	248	252	236	250	174	190	239	255
EM	34	190	210	240	244	268	268	190	208	253	255
	333	238	250	235	235	239	257	194	220	248	260
Fercal		184	190	220	244	244	244	184	220	245	257
Freedom		178	187	248	248			188	190	257	267
Golia		208	208	220	258			190	196	255	263
Gravesac		188	212	240	240	266	266	190	192	255	257
Grézot	1	180	212	242	244	238	238	188	191	255	261
Harmony		210	214	242	248			188	202	255	257
Kober	5BB	192	212	216	252	260	260	200	214	251	259
	125AA	204	212	216	252	260	260	192	210	251	259
Kraus1*		183	210					179	199	253	261
LN 33		178	195	244	248	250	250	188	190	259	267
Malégue	44–53	206	216	228	252	250	250	174	178	257	261
Mgt	101–14	198	208	242	250	236	236	172	190	255	257
	420A	194	218	244	246	264	264	190	196	255	255

(continued)

Table 10.2 (continued)

Rootstock	Clones	VVMD27		VVMD28		VVMD32		VRZAG62		VRZAG79	
		A1	A2	A1	A2	A1	A2	A1	A2	A1	A2
Mgt	41B	190	192	242	268	240	256	194	220	255	259
NAB		195	206			250	252	214	214	257	259
NOAH		186	188					181	206	249	259
Paulsen	775	192	208	220	252	260	260	196	210	251	263
	779	204	206	242	252	260	260	196	210	251	259
	1045	190	204	242	252			196	214	251	259
	1103	204	208	242	252	260	260	196	214	251	263
	1447	204	208	220	252	260	260	196	214	251	263
Ramsey		192	214	220	242	237	237	174	202	237	255
Richter	31	188	188	220	244			184	190	257	259
	57			234	240			196	214	243	263
	99	192	208	220	236	260	260	196	210	251	263
	110	190	206	220	234	254	254	196	214	243	259
Rici		186	208	242	260	256	256	174	190	251	255
RGM		208	212	216	244	237	237	192	200	255	259
R.S.B	1	236	246	235	243	259	259	200	214	252	260
Ruggeri	140	190	206	234	242	252	252	196	214	243	259

(continued)

Table 10.2 (continued)

Rootstock	Clones	VVMD27		VVMD28		VVMD32		VRZAG62		VRZAG79	
		A1	A2	A1	A2	A1	A2	A1	A2	A1	A2
Rupestris du lot		206	208	220	242	235	237	196	196	259	263
Schwarzmann		208	214	216	244	235	237	200	202	255	257
SO4		238	249	214	235	259	259	200	214	252	256
Sori		210	218	244	248	236	264	190	190	259	261
Sorisil		195	218	236	248	272	272	180	190	249	259
Teleki	8B	204	212	216	252	260	260	192	214	251	259
	5C	204	212	216	252	260	260	200	210	251	259
Vialla		188	210	226	236	246	246	202	202	249	255

References

Al Rwahnih M, Daubert S, Golino D, Islas C, Rowhani A (2015) Comparison of next generation sequencing versus biological indexing for the optimal detection of viral pathogens in grapevine. Phytopathology 105(6):758–763

Alaniz S, Armengol J, León M, García-Jiménez J, Abad-Campos P (2009) Analysis of genetic and virulence diversity of *Cylindrocarpon liriodendri* and *C. macrodidymum* associated with black foot disease of grapevine. Mycol Res, Cambridge, 113(1):16–23

Bacilieri R, This P (2010) GrapeGen06, an European project for the management and conservation of grapevine genetic resources. http://www1.montpellier.inra.fr/grapegen06/accueil.php

Bianchi D, Brancadoro L, Lorenzis G DE (2020) Genetic diversity and population structure in a *Vitis* spp. core collection investigated by SNP markers. Diversity 12(3, Art. 103). https://doi.org/10.3390/d12030103

Bisztray GD, Korbuly J, Halász J, Oláh R, Sz R, Deák T, Pedryc A (2003) Characterization of grape varieties and species by RAPD markers. Acta Hortic 603:601–604

Bourquin JC, Otten L, Walter B (1991) Identification of grapevine rootstocks by RFLP. Comptes Rendus De L Academie Des Sciences Serie Iii-Sciences De La Vie-Life Sciences 312(12):593–598

Bourquin JC, Otten L, Walter B (1995) PCR-RFLP analysis of Vitis, Ampelopsis and parthenocissus and its application to the identification of rootstocks. Vitis 34(2):103–108

Bourquin JC, Tournier P, Otten L, Walter B (1992) Identification of sixteen grapevine rootstocks by RFLP and RFLP analysis of nuclear DNA extracted from the wood. Vitis 31(3):157–162

Boursiquot JM, Parra P (1992) Application d'une methode d'electrophorese pour la caracterisation et al reconnaissance des porte- greffe. Vitis 31:189–194

Bowers JE, Dangl GS, Vignani R, Meredith CP (1996) Isolation and characterization of new polymorphic simple sequence repeat loci in grape (*Vitis vinifera* L.). Genome 39:628–633

Cabezas JA, Ibáñez J, Lijavetzky D, Vélez D, Bravo G, Rodríguez V, Carreño I, Jermakow AM, Carreño J, Ruiz-García L, Thomas MR, Martinez-Zapater JM (2011) A 48 SNP set for grapevine cultivar identification. BMC Plant Biol 11(1):1–12

Crespan M, Meneghetti S, Cancellier S (2009) Identification and genetic relationship of the principal rootstocks cultivated in Italy. Am J Enol Vitic 60(3):349–356

Dangl GS, Mendum ML, Prins BH, Walker MA, Meredith CP, Simon CJ (2001) Simple sequence repeat analysis of a clonally propagated species: a tool for managing a grape germplasm collection. Genome 44:432–438

de Andrés MT, Cabezas JA, Cervera MT, Borrego J, Martínez-Zapater JM, Jouve N (2007) Molecular characterization of grapevine rootstocks maintained in germplasm collections. Am J Enol Vitic 58:75–86

Di Gaspero G, Peterlunger E, Testolin R, Edwards KJ, Cipriani G (2000) Conservation of microsatellite loci within the genus Vitis. Theor Appl Genet 101:301–308

D'onofrio C (2020) Introgression among cultivated and wild grapevine in Tuscany. Frontiers Plant Sci 11(Art. 202). (https://doi.org/10.3389/fpls.2020.00202)

EPPO, European and Mediterranean Plant Protection Organization (2008) Certification scheme, Pathogen-tested material of grapevine varieties and rootstocks. Bulletin EPPO Bulletin 38:422–429

Fan XC, Chu JQ, Liu1 CH, Sun X, Fang JG (2014) Identification of grapevine rootstock cultivars using expressed sequence tag-simple sequence repeats. Genet Mol Res 13(3):7649–7657

Fiore N (2015) Enfermedades de la vid causadas por virus, viroides y fitoplasmas: diagnóstico, epidemiologíay control. Congresso Latino-Americano de Viticultura e Enologia, Bento Gonçalves, RS. Resumos, Bento Gonçalves: Embrapa Uva e Vinho, pp. 118-130

Golino DA (2000) Trade in grapevine plant materials: local, national, and worldwide perspectives. In: Proceedings of the ASEV 50th anniversary meeting, Seattle, Washington June 19–23, 2000, pp 216–221

Gramaje D, di Marco S (2015) Identifying practices likely to have impacts on grapevine trunk disease infections: a European nursery survey. Phytopathologia Mediterranea, Bologna 54(2):313–324

Grohs DS, Almanca MAK, Fajardo TVM, Alleen FH, Miele A (2017) Advances in propagation of grapevine in the world. Rev Bras Frutic 39(4):e-760. https://doi.org/10.1590/0100-29452017760

Halász J, Korbuly J, Deák T, Bisztray GD (2004) RAPD analysis of grapevine hybrids and cultivars. Int J Hortic Sci 10(4):63–66

Jahnke G, Molnár GK, Májer J, Szőke B, Tarczal E, Varga P, Kocsis L (2011) Analysis of grape rootstocks by SSR markers. J Int Des Sci De La Vigne Et Du Vin 45(3):1–12

Laucou V, Boursiquot JM, Lacombe T, Bordenave L, Decroocq S, Ollat N (2008) Parentage of grapevine rootstock 'Fercal' finally elucidated. Vitis 47(3):163–167

Laucou V, Launay A, Bacilieri R, Lacombe T, Adam-Blondon AF, Bérard A, Chauveau A, Teresa de Andrés M, Hausmann L, Ibáñez J, Le Paslier MC, Maghradze D, Martinez-Zapater JM, Maul E, Ponnaiah M, Töpfer R, Péros JP, Boursiquot JM (2018). Extended diversity analysis of cultivated grapevine *Vitis vinifera* with 10K genome-wide SNPs. PLoS ONE 13(2):e0192540

Lázár J (2003) Sanitary aspects and results of the Hungarian grape breeding. Acta Hortic 603:755–762

Leão PCS, Riaz S, Graziani R, Dangl GS, Motoike SY, Walker MA (2009) Characterization of a Brazilian grape germplasm collection using microsatellite markers. Am J Enol Vitic 60(4):517–524

Lin H, Walker MA (1997) Extracting DNA from cambium tissue for analysis of grape rootstocks. HortScience 32:1264–1266

Lin H, Walker MA (1998) Identifying grape rootstocks with simple sequence repeat (SSR) DNA markers. Am J Enol Vitic 49(4):403–407

Lowe KM, Walker MA (2006) Genetic linkage map of the interspecific grape rootstock cross Ramsey (*Vitis champinii*) × Riparia Gloire (*Vitis riparia*). Theor Appl Genet 112:1582–1592

Maliogka VI, Skiada FG, Eleftheriou EP, Katis NI (2009) Elimination of a new Ampelovirus (GLRaV-Pr) and grapevine rupestris stem pitting associated virus (GRSPaV) from two *Vitis vinifera* cultivars combining in vitro thermotherapy with shoot tip culture. Scientia Horticulturae, Amsterdam 123(2):280–282

Martelli GP (2014) Directory of virus and virus-like diseases of the grapevine and their agents. J Plant Pathol Bari 96(1):1–136

Martelli GP, Saldarelli P (2015) Phytosanitary challenges for the Mediterranean viticultural industry: emerging grapevine viruses. CIHEAM, International Centre for Advanced Mediterranean Agronomic Studies, Bari, 4 p

Maul E, Töpfer R (2015) Vitis international variety catalogue (VIVC): a cultivar database referenced by genetic profiles and morphology. In: BIO web of conferences (5:01009). EDP Sciences. https://doi.org/10.1051/bioconf/20150501009

Miller AJ, Matasci N, Schwaninger H, Aradhya MK, Prins B, Zhong GY, Simon C, Buckler ES, Myles S (2013) Vitis phylogenomics: hybridization intensities from a SNP array outperform genotype calls. PLoS ONE 8(11):e78680. https://doi.org/10.1371/journal.pone.0078680.g001

Moravcova K, Baranek M, Pidra M (2006) Use of SSR markers to identify grapevine cultivars registered in the Czech Republic. J Int Vigne Vin 40(2):71–80

Rodrigues Neto J, Destéfano SAL, Rodrigues LMR, Pelloso DS, da C. Oliviera Júnior L (2011) Grapevine bacterial canker in the State of São Paulo, Brazil: detection and eradication. Trop Plant Pathol Brasília, DF 36(1):42–44

Schmid J, Manty F, Rühl EH (2015) Optimisation of sustainability of grapevine varieties by selecting rootstock varieties under different environmental conditions and creating new rootstock varieties. http://www.viticulturasustentavel.com/pdfs/Prof_Joachim_Schmid.pdf. Accessed on 5 Oct 2019

Sefc KM, Regner F, Turetschek E, Glössl J, Steinkellner H (1999) Identification of microsatellite sequences in *Vitis riparia* and their applicability for genotyping of different Vitis species. Genome 42:367–373

Stamp JA (2001) The contribution of imperfections in nursery stock to the decline of young vines in California. Phytopathologia Mediterranea, Bologn 40(3):369–375

This P, Cuisset C, Boursiquot JM (1997) Development of stable RAPD markers for the identification of grapevine rootstocks and the analysis of genetic relationships. Am J Enol Vitic 48(4):492–501

This P, Lacombe T, Thomas MR (2006) Historical origins and genetic diversity of wine grapes. Trends Genet 22:511–519

Thomas MR, Cain P, Scott KD (1994) DNA typing of grapevines: a universal methodology and database for describing cultivars and evaluating genetic relatedness. Plant Mol Biol 25:939–949

Thomas MR, Scott NS (1993) Microsatellite repeat in grapevine reveal DNA polymorphisms when analysis as sequence-tagged sites (STSs). Theor Appl Genet 86:985–990

Tsai CW, Rowhani A, Golino DA, Daane KM, Almeida RP (2010) Mealybug transmission of Grapevine leafroll viruses: an analysis of virus–vector specificity. Phytopathol Saint Paul 100(8):830–834

Upadhyay A, Saboji MD, Reddy S, Deokar K, Karibasappa GS (2007) AFLP and SSR marker analysis of grape rootstocks in Indian grape germplasm. Sci Hortic 112(2):176–183

VIVC, Vitis International Variety Catalogue (2019) Passbord data, Utilization: Rootstock. http://www.vivc.de

VIVC, Vitis International Variety Catalogue (2021) Single nucleotide polymorphisms (SNPs) by varieties. https://www.vivc.de/index.php?r=snp112ptrans%2Fresult2

Chapter 11
Commercial Grape Rootstocks Selections

Around 1500 grape rootstocks have been selected or bred in the world to date (VIVC 2019). Of those, only around 50 of them are common in commercial use. However, in the following section, around 150 rootstocks are described. The seed parent (maternal) and the pollen parent (paternal) are listed respectively for the hybrid rootstocks, as shown in each pedigree. The rootstocks are organized by their species. The descriptions, especially the resistance to pests, should be interpreted carefully as different locations, growing conditions, desired viticultural results, and even the specific population of a pathogen can affect the responses that are measured and, consequently, the selection of rootstock to be used. There are efforts to study the evolution of the rootstock agro-eco-system with the changing climate and pests and disease pressure. Although there is no clear answer, the overall trend is for an increased pest and disease pressure that will make their control more difficult. In any case, viticulture will face a more complicated situation that requires production to be based on a more resilient system. Therefore, this information will help growers to make better decisions (Micheloni 2017). The shapes of the leaves of the most important rootstocks are also provided in Figs. 11.1, 11.2, 11.3, 11.4, 11.5, 11.6, 11.7 and 11.8.

A. Rahemi et al., *Grape Rootstocks and Related Species*,
https://doi.org/10.1007/978-3-030-99407-5_11

Fig. 11.1 Blade shapes of mature leaves of commercial grape rootstocks (References for this figure is available at the end of this chapter)

Fig. 11.2 Blade shapes of mature leaves of commercial grape rootstocks (References for this figure is available at the end of this chapter)

Fig. 11.3 Blade shapes of mature leaves of commercial grape rootstocks (References for this figure is available at the end of this chapter)

Fig. 11.4 Blade shapes of mature leaves of commercial grape rootstocks (References for this figure is available at the end of this chapter)

Fig. 11.5 Blade shapes of mature leaves of commercial grape rootstocks (References for this figure is available at the end of this chapter)

Fig. 11.6 Blade shapes of mature leaves of commercial grape rootstocks (References for this figure is available at the end of this chapter)

Fig. 11.7 Blade shapes of mature leaves of commercial grape rootstocks (References for this figure is available at the end of this chapter)

Fig. 11.8 Blade shapes of mature leaves of commercial grape rootstocks (References for this chapter is available at the end of this chapter)

11.1 *Vitis acerifolia*

V. acerifolia includes two separate species, *V. solonis* and *V. longii* (Laguna Lumbreras 2004a). The rootstocks with the pedigree of *V. solonis* and *V. longii* are presented separately below.

11.2 *Vitis aestivalis*

11.2.1 *V. aestivalis* × *V. labrusca* × *V. labrusca* × *V. vinifera* × *V. labrusca* × *V. labrusca*

(a)   **Tampa** (Florida 43-47 (open-pollinated *V. aestivalis* 'Smalliana Bailey') × 'Niagara' (*V. labrusca* 'Concord' ('Catawba' (*V. labrusca* × *V. vinifera* 'Semillon') × *V. labrusca*) × 'Cassady'(*V. labrusca*)))

This rootstock is recommended in several areas of Florida, which have Pierce's disease problem. It is also shown to have a root-knot nematode tolerance. Tampa has a large trunk and cane framework with vigorous growth (Mortensen and Stover 1982).

11.2.2 *V. aestivalis* × *V. labrusca* × *V. vinifera* × *V. labrusca*

(a)   **Blue Lake** (Florida 43-47 (open-pollinated of *V. aestivalis* 'Smalliana') × 'Caco' ('Catawba' (*V. labrusca* × *V. vinifera* 'Semillon') × 'Concord' (*V. labrusca*))

Blue Lake is another rootstock recommended for areas with Pierce's disease pressure in Florida (Mortensen and Stover 1982). It can also be used as an own-rooted bunch grape (Stover 1960).

11.2.3 *V. aestivalis* × *V. monticola* × *V. riparia* × *V. rupestris*

(a)   **554-5 C (554–5 Couderc)** (*V. aestivalis* × *V. monticola*) × (*V. riparia* × *V. rupestris*)

554-5 is susceptible to root-knot nematodes (Clingeleffer and Davis 2012) but has demonstrated good tolerance to grapevine leaf rust (Daly and Hennessy 2007).

11.2.4   *V. aestivalis* × *V. vinifera*

(a)   **Jacquez** (a natural crossing of *V. aestivalis* × *V. vinifera*)

Jacquez is an own-root grape that originated from Portugal and moved to South Africa and was used as a rootstock in the late nineteenth century. It shows a good performance in heavy loam and clay soils and in subsoil wetness conditions (Teubes 2014).

11.2.5   *Vitis amurensis*

V. amurensis usesin modern breeding rootstock (Alleweldt and Possingham 1988). Amos is one of the rootstocks which has *V. amurensis* in its genetic background. For its characteristics, see *V. vinifera* below.

11.3   *Vitis berlandieri*

11.3.1   *V. berlandieri*

(a)   **Berlandieri 'Rességuier 1' (Rességuier n°1) and Berlandieri 'Rességuier 2' (Rességuier n°2)** (*V. berlandieri*)

These two small-leafed drought-tolerant rootstocks are both considered to be *V. berlandieri* which originated from hillsides and were used as *V. berlandieri* parent species in many breeding programs (Teubes 2014). Rességuier1 has all the characteristics required in a rootstock, such as a phylloxera resistance, ease to graft, permanency of union, fertility of the scion, adaptation to varied soils, but it has a weakness which not easy to root (Bioletti 1908).

11.3.2   *V. berlandieri* × *V. berlandieri*

(a)   **Hochberg 48 (H48)** (*V. berlandieri* 'Dabouki' × *V. berlandieri* 'Planchon')

Hochberg 48 originated from Israel and was introduced to the US as a rootstock for high-yielding varieties (FPS Grapes 2018).

(b) **M 5489 (Merbein 5489)** (*V. berlandieri* 'Rességuier 1' × *V. berlandieri* 'Mazade' (syn. *V. cinerea* var. Helleri 'Rességuier 1' × *V. cinerea* var. Helleri 'Mazade'))

It is an Australian rootstock with a medium vigor size. M4589 in Australia showed a similar yield with Ramsey and 1103 Paulsen while with half the pruning weight. M4589 showed resistance to some types of phylloxera. Using M4589 in nematode-infested fields showed no losses in growth and yield. It has similar sodium and chloride uptake with 1103P and lower juice potassium than 1103P. 'Shiraz' grafted on M5489 showed higher color and total phenolics of wine compared with those which grafted on 1103P or Ramsey (Walker et al. 2014).

(c) **M 5512 (Merbein 5512)** (*V. berlandieri* 'Rességuier 1' × *V. berlandieri* 7651 (syn. *V. cinerea* var. Helleri 'Rességuier 1' × *V. cinerea* var. Helleri 7651))

The M5512 rootstock is Australian in origin and has a medium vigor. M5512 has about one-third the pruning weight of Ramsey and 1103P in Australia. M5512, when compared with 1103P and Ramsey, showed 25% and 47% yield reductions, respectively. M5512 has resistance to some strains of phylloxera. M5512 in nematode-infested vineyards showed no suffering effects during field trials. It showed a similar uptake of both chloride and sodium as 1103P and Ramsey. M5512 showed less potassium content in the berries as compared to 1103P and Ramsey (Walker et al. 2014).

11.3.3   *V. berlandieri* × *V. longii*

(d) **31 R (31 Richter)** (*V. berlandieri* 'Rességuier 2' × *V. longii* 'Novo-mexicana')

31R has foliar morphology very similar to 99R and 110R (Laguna Lumbreras 2004b). It has shown a susceptibility to *Meloidogyne javanica* nematode (Clingeleffer and Smith 2011). It is considered a likely parent of the Fercal rootstock (Laucou et al. 2008).

11.3.4   *V. berlandieri* × *V. riparia*

The crosses between these two species are made the largest and important group of rootstocks used worldwide. The *V. riparia* background brings easiness of propagation and grafting and tolerance to wet and cool conditions. The *V. berlandieri* background brings tolerance to drought and lime conditions. These rootstocks usually show good vigor if they receive enough water and generally grow in northern areas (Walker 1991).

(a) **5A (5A, Teleki 5A, 5A-Teleki)** (*V. berlandieri* × *V. riparia*)

Thought to be the same seed lot, or possibly a parent of Kober 5BB (Galet 2004). The genetic results show that these are the same rootstock. Although later, when 5A Teleki and Kober 5BB were compared, the trials suggested it be considered a clonal comparison (Whiting 2003).

(b) **5BB (5BB, Kober 5BB)** (*V. berlandieri* × *V. riparia*)

It has a good phylloxera resistance (Galet 2004; Walker 1991; INRA 2016). It is reported that it has some tolerance (Galet 2004; Pongrácz 1983) to moderate resistance (Walker 1991) to nematodes. It has a high tolerance to *Meloidogyne incognita* and *M. hapla*, and moderate resistance to *M. arenaria* (INRA 2016). It is shown to be susceptible to *Agrobacterium vitis*, highly resistant to downy mildew (INRA 2016), and susceptible to phytophthora (Bettiga et al. 2003). Some graft union issues have been reported when grafted to some cultivars infected with leafroll virus 2 (INRA 2016). It has moderate tolerance to drought and salinity and moderate to the high tolerance of lime. Its nutrient profile shows that it has a moderate to high ability to uptake nitrogen, calcium, and magnesium and a moderate ability to uptake phosphorus, potassium, and zink (Bettiga et al. 2003). It is considered to be well adapted to moist clay soils. In France, it is considered best adapted to heavy clay soils and can produce too vigorous vines (Galet 2004). In Europe, it is considered to have substantial vigor, with an ability to withstand up to 35% limestone in the soil. In sandy sites, the 5BB is suggested for humid conditions (INRA 2016). In California, the 5BB is considered a moderate vigor rootstock (Walker 1991). In South Africa, it is also considered a good rootstock for humid sites with compact calcareous clay soils and especially recommended for areas that require early ripening (Pongrácz 1983). 'Chardonnay' and 'Cabernet franc' grapevines grafted on 5BB rootstock produced greater yields, higher pruning weights, and lower °Brix compared to Riparia in Canada (Vanden Heuvel et al. 2004).

(c) **5C (Teleki 5C, 5C Geisenheim, 5C)** (*V. berlandieri* × *V. riparia*)

The 5C shows a good (INRA 2016) to strong (Walker 1991) and high (Bettiga et al. 2003) phylloxera resistance. 5C generally has moderate nematode resistance (Walker 1991), with medium to high resistance to root-knot nematodes (Bettiga et al. 2003). It is very resistant to *Meloidogyne incognita* (INRA 2016), but it is low to moderately resistant to the *Xiphinema index* (Bettiga et al. 2003). It is also highly resistant to downy mildew (INRA 2016). In California, 5C has shown a low to moderate vigor rootstock. It has a low tolerance to drought, low to moderate tolerance to wet soil but is adapted to moist clay soils. It also has moderate tolerance to salinity. 5C has a low ability to uptake nitrogen, low to moderate ability to uptake zinc, moderate ability to uptake potassium and phosphorus, and moderate to high ability to uptake magnesium from the soil (Bettiga et al. 2003). 5C is reported to beis not well suited to warm irrigated areas and may have a low juice pH (Walker 1991). In Europe, 5C is considered to have substantial vigor, medium to good lime tolerance, and good quality crop when used as a rootstock for 'Riesling' (INRA 2016). Jahnke et al.

(2011) reported that SSR markers showed a great difference between the 5C and 5BB and suggested there are different genotypes with highly similar morphological features.

(d)   **8B (8B, Teleki 8B, Crâciunel 71, C-71)** (*V. berlandieri* × *V. riparia*)

The 8B is highly tolerant to radicicola phylloxera, and very highly resistant to the root-knot nematodes, *Meloidogyne arenaria,* and *M. incognita.* It also has high resistance to downy mildew. It is considered as a moderate vigor, and resistance to limestone, and good resistance to drought in Australia (INRA 2016).

(e)   **34EM (34 Ecole De Montpellier)** (*V. berlandieri* Ecole × *V. riparia*)

Rootstock 34EM has strong resistance to radicicola phylloxera, and is resistant to *Meloidogyne incognita,* but only moderately resistant to *M. arenaria.* It is also shown high resistance to downy mildew and moderate tolerance to lime. In Europe itis thought to hasten scion maturity (INRA 2016).

(f)   **125AA (Kober 125AA, 125AA)** (*V. berlandieri* × *V. riparia*)

125AA is considered to have good (Galet 2004) to high tolerance (INRA 2016) to radicicola phylloxera. It is also highly resistant to downy mildew (INRA 2016). 125AA shows moderate lime tolerance and is commonly grown in heavier wet soils in France (Galet 2004). It is very easy to root, and growth is generally vigorous. The production of wood is average to good. Grafted grapes on this rootstock also show vigorous to moderately vigorous growth. It has good resistance to winter weather (−27 to −30 °C). However, the root system is sensitive to winter damage. 125AA is minimally resistant to drought (medium to low). The rootstock is asuitable counter-part for very fertile scion varieties. There is no need for large amounts of nitrogen fertilizer, and it can spread the doses, as this suppresses potassium intake. It is reported for its difficulty in absorbing magnesium from the soil in France (INRA 2016), but it is considered to easily absorb magnesium in the Czech (Šimek 2017). It has an average resistance to the high calcium content in the soil and can tolerate 13–18% of active Calcium in the soil. It shows slightly below average to average resistance to ferric chlorosis. 125AA is suitable for worse soil conditions in loamy and sandy-loam, sufficiently moist soils and is not suggested for shallow, dry, and too rocky soils. 125AA can also withstand the heavier, marshy clay soils. However, it is more sensitive to excessive soil moisture in spring. It has a good affinity with varieties of *Vitis vinifera* (INRA 2016; Šimek 2017).

(g)   **157-11C (157-11 Couderc)** (*V. berlandieri* Las Sorres × *V. riparia* 'Gloire de Montpellier')

157-11C has the leaf shape same as *V. berlandieri*, and a root system intermediate between its parents. Resistant to phylloxera, grafted scion shows vigorous growth, high yield, ripen well, with easy to root and graft (Bioletti 1908).

(h) **161-49C (161-49 Couderc)** (*V. berlandieri* × *V. riparia*)

161-49C has a high tolerance to radicicola phylloxera, but is moderately susceptible to foliar phylloxera. It is moderate resistant to *Meloidogyne incognita* nematodes and susceptible to *Meloidogyne arenaria*. 161-49C also has a high resistance to downy mildew. 161-49C is well adapted to limestone and clay-limestone soils which are not very compact. This rootstock has moderate to good tolerance to drought but especially when it is young, it shows leaf wilting (tylosis which sap becomes blocked). This rootstock is also susceptible to temporary excess humidity in the spring. Some cases of dieback of grafted vines on 161-49C have been reported especially in the Southern US, but some cases are also reported in other areas. Initial symptoms (substantial decrease in vigor) often occur from the third year after planting, and cambium failure has been seen in affected vines with an unusual trunk and root thickening, which has appeared under soil and climate conditions. The growth and time to fruiting of scion are fairly slow. It shows good quality crops and good affinity grafts with 'Chardonnay', 'Colombard', 'Merlot', 'Pinot noir', and 'Ugni blanc'. It also produces good results with interspecific hybrid cultivars, but some grafting problems are reported with 'Carignan', 'Gamay', and 'Servant'. 161-49C is a moderate vigor rootstock with a moderate wood producer and sometimes a certain proportion of damaged wood. The cutting and grafting capacity of 161-49C is moderate and needs special attention in the nursery (INRA 2016).

(i) **225 Ru (225 Ruggeri)** (*V. berlandieri* × *V. riparia*)

The 225Ru rootstock has a tolerance to phylloxera and is moderately tolerant to wet soil and drought conditions. It is easy to root and graft (Elkins and Walker 1991).

(j) **420A (420A-Mgt, Millardet et de Grasset 420A, 420A)** (*V. berlandieri* × *V. riparia* 420A)

The 420A is a low vigor rootstock that induces a slightly earlier maturity, and wood maturity in comparison to Riparia Gloire. It is not tolerant to drought (Galet 2004) or water-logged soils. However, it can grow well in heavy loam and clay soils (Pongrácz 1983; Whiting and Buchanan 1992). It shows weak growth and low yields in Australia (Walker 1991). 420A is highly tolerant to radicicola phylloxera but moderate to foliar phylloxera. It has resistance to the nematodes *Meloidogyne incognita* and *Meloidogyne arenaria*, and it shows good resistance to downy mildew (INRA 2016). 420A resists up to 35% total limestone, 20% active limestone, and a CPI of 40. It has moderate to good resistance to ferric chlorosis. It is well adapted to fertile soils and to quite deep clay-limestone soils with adequate water. It is not well adapted to compact soils and excessive spring humidity in the soil. 420A rootstock has difficulty absorbing potassium, and grafted grape cultivars can frequently show potassium deficiency with this rootstock. The grafted vine shows a low vigor on this rootstock and with low trunk development. 420A tends to encourage floral initiation but shows a delay in maturity. Due to the vigor of the grapevines on this rootstock, production yield should be limited, especially in the early years of production, to avoid unbalanced trunks. 'Cabernet franc', 'Merlot', and 'Tannat' produce good results with

420A rootstock. Wood production is moderate with the cane lignifyingeasily, but the diameter of the cane is typically rather thin. There is a considerable growth of quick buding. The cutting potential is low, and there is a need to apply rooting hormone. The grafting capacity is moderate.

(k) **Crâciunel 2 (C-2)** (*V. berlandieri* × *V. riparia*)

Crâciunel 2 originated from the selection of Kober 5BB rootstock (VIVC 2019). The vegetation cycle of the variety from sprouting to leaf fall lasts 180 days (6–8 days is shorter than 5BB) at a sum of active temperatures (SAT) of 3250 °C. Crâciunel 2 is vigorous, with very good rooting. Winter resistance is about −27 °C for the root system. It has good resistance to radicicola phylloxera and to ferric chlorosis and is resistant to downy mildew (*Plasmopara viticola*) and powdery mildew (*Uncinula necator*). Its drought resistance is below average but better than the 5BB. It is resistant to the soil calcium content (up to approximately 20% of active Ca), which is better than 5BB. It is suggested for medium fertile soils, loess, sandy, and gravel soils, but not good for heavy and compact soils, high salt content, and acid soils. Crâciunel 2 has a good graft affinity with *Vitis vinifera* cultivars (Šimek 2017).

(l) **RSB 1 (Rességuier Selection Birolleau 1, R.S.B. 1)** (*V. berlandieri* × *V. riparia*, derived from Euryale Rességuier)

RSB1 has a high tolerance to radicicola phylloxera and low to moderate susceptibility to gallicola phylloxera. It displays high resistance to downy mildew and anthracnose. It can resist up to 40% total limestone, 20% active limestone, and a CPI of 50. It has good resistance to ferric chlorosis but cannot absorb magnesium well. This rootstock is adapted to not very fertile soils and temporary excess humidity in the spring condition, and it shows good resistance to summer drought. RSB1 also shows a good graft affinity (clone# 109 of RSB1 sometimes shows problems with grafting). It has strong vigor with quite a high yield. 'Ugni blanc' displays a very good yield with RSB1. This rootstock is a very good wood producer, and it shows easy of cutting and grafting. The cane is long with moderate to large diameter (INRA 2016).

(m) **SO4 (SO4, Selektion Oppenheim 4, Oppenheim #4)** (from Teleki 4A (*V. berlandieri* 'Rességuier' × *V. riparia*))

The SO4 is highly resistant to radicicola phylloxera and has a low to medium susceptibility to gallicola phylloxera and anthracnose, and shows high resistance to downy mildew. It has very good resistance to the nematodes *Meloidogyne incognita* and *Meloidogyne arenaria*, and moderate resistance to *Meloidogyne hapla*. This rootstock sometimes shows susceptibility to Dutch elm disease. (INRA 2016). It seems this rootstock affects scion varieties medium to short-cycling on fruit and canopy maturation period (Howell 2005; Perry and Sabbatini 2015), and in suitable soils, it ensures a good set and advances maturity. It has good lime tolerance (Pongrácz 1983). SO4 resists up to 35% total limestone, 17% active limestone, and a CPI of 30. It has moderate resistance to ferric chlorosis, and it is well suited to acidic soils (INRA 2016). It has medium to good resistance to drought, but its adaptation to humidity is low to medium. The SO4 cannot absorb magnesium very well and creates a black stem in the vine. This rootstock is well suited to sandy soils and clay-limestone soils, which are moderately or not fertile. SO4 has a good affinity of grafts, but the trunk radial growth is very limited and shows differences in diameter with the scion. The grafted varieties on SO4 show vigor in growth, and it is a good wood producer, but wood lignifies fairly late. This rootstock is easy to propagate and to graft. Its canes have a medium-sized diameter, and the growth of quick buds is limited. This rootstock has a high crop yield with good sugar contents and produces wines with high pH levels (INRA 2016).

(n)   **Binova (Selektion Oppenheim 4 Mutation)** (*V. berlandieri* × *V. riparia*)

It is a sport of SO4. While the SO4 has male flowers, Binova has hermaphrodite flowers. Binova has similar traits to SO4 and even shows more limestone tolerance (up to 50% total limestone, 30% active limestone) (Ruehl et al. 2015; Wine growers 2019).

(o)   **Cosmo 2 (Teleki Cosmo 2)** (*V. berlandieri* × *V. riparia*)

Cosmo2 is selected from Teleki 8B at an Experimental Station for Viticulture at Conegliano (Bavaresco et al. 2015). It is resistant to *Meloidogyne javanica* (Clingeleffer and Smith 2011) and has medium tolerance to ferric chlorosis and drought (Pavloušek 2013).

(p)   **Cosmo 10 (Teleki Cosmo 10)** (*V. berlandieri* × *V. riparia*)

Cosmo10 is also selected from Teleki 8B at the Experimental Station for Viticulture at Conegliano (Bavaresco et al. 2015), but it is susceptible to *Meloidogyne javanica* (Clingeleffer and Smith 2011) and has medium tolerance to ferric chlorosis and drought (Pavloušek 2013).

11.3.5   *V. berlandieri* × *V. riparia* × *V. berlandieri* × *V. riparia*

(a)   **M 3 (M3)** (R 27 (*V. berlandieri* × *V. riparia*) × Teleki 5C (*V. berlandieri* × *V. riparia*))

M3 was created by the University of Milan, Italy. It is a low vigor rootstock with high capability in potassium uptake. M3 has low resistance to salt (Bavaresco et al. 2015).

(b)   **Star 50** (Binova (*V. berlandieri* × *V. riparia*) × Binova (*V. berlandieri* × *V. riparia*))

It is derived from self-pollinated Binova in a breeding program at the University of Bologna, Italy. It is a low vigor rootstock and has good resistance to phylloxera and lime-induced chlorosis (Bavaresco et al. 2015).

(c)   **Star 74** (Binova (*V. berlandieri* × *V. riparia*) × Binova (*V. berlandieri* × *V. riparia*))

It is also obtained from self-pollinated Binova in a breeding program at the University of Bologna, Italy. It is low vigor and shows good resistance to phylloxera and lime-induced chlorosis (Bavaresco et al. 2015).

11.3.6 *V. berlandieri* × *V. riparia* × *V. cinerea*

(a)   **Cina** (125AA (*V. berlandieri* × *V. riparia*) × *V. cinerea* Arnold)

It has medium to high drought resistance and low to medium chlorosis resistance, which is bred in Germany (Pavloušek 2013).

11.3.7 *V. berlandieri* × *V. riparia* × *V. riparia* × *V. rupestris*

(a)   **Gravesac** (161-49 Couderc (*V. berlandieri* × *V. riparia*) × 3309 Couderc (*V. riparia* Tomenteuse × *V. rupestris* Martin))

Gravesac has a high tolerance level to radicicola phylloxera and is moderately sensitive to gallicola phylloxera. It shows good resistance to downy mildew and anthracnose, but it is sensitive to the nematodes *Meloidogyne incognita* and *Meloidogyne arenaria*. It is adapted to sandy or gravel soils. Gravesac resists up to 15% total limestone, 6% active limestone, and a CPI of 5. It has a low to moderate resistance to ferric chlorosis. It displays very good behavior on acidic soils. This rootstock is also adapted to temporary humid conditions in spring, and it has moderate resistance to drought. Gravesac shows a good graft affinity and has a moderate to strong vigor. The grafted vines on the Gravesac show a high and stable yield with good quality. Gravesac is a good producer of hard wood with moderate to large diameter canes, with easy of cutting and graftability (INRA 2016).

11.3.8 *V. berlandieri* × *V. riparia* × *V. vinifera*

(a)   **Georgikon 28** (Kober 5 BB (*V. berlandieri* × *V. riparia*) × *V. vinifera*)

Georgikon28 has an excellent phylloxera resistance, better than 5C (Tóth and Kocsis 2003), and has a high tolerance to drought and chlorosis (Pavloušek 2013).

11.3.9 *V. berlandieri* × *V. riparia* × *V. vinifera* × *V. berlandieri*

(a) **M 2 (M2)** (Teleki 8B (*V. berlandieri* × *V. riparia*) × 333 E.M. (*V. vinifera* × *V. berlandieri*)

M2 is bred by the University of Milan, Italy. It shows a medium vigor, and has a good resistance to ferric chlorosis, and medium resistance to salt (Bavaresco et al. 2015).

11.3.10 *V. berlandieri* × *V. rupestris*

These hybrids are crossed between an easily propagated species (*V. berlandieri*) and a difficult to root species (*V. rupestris*), but both are adapted to drought stress, and rootstocks produced from them are adapted to sites where water is limited. These root-stocks produce shoots with greater numbers of laterals and reduced overall growth compared to the *V. berlandieri* × *V. riparia* rootstocks. In general, the *V. berlandieri* × *V. rupestris* rootstocks were designed for southern European, Mediterranean growing conditions in non-irrigated vineyards (Walker 1991).

(a) **57R (57 Richter)** (*V. berlandieri* 'Rességuier 1' × *V. rupestris* Martin)

57R has shown a resistance to *Meloidogyne javanica* (Clingeleffer and Smith 2011). It is reported that the 'Emperor', 'Molinera', 'Palomino', 'Aramon' (Read and Gu 2003), and 'Jaoumet B' (Plantgrape 2019) are incompatible with 57R.

(b) **99 R (99 Richter, Prosperi Super 99 R)** (*V. berlandieri* Las Sorres × *V. rupestris* 'du Lot')

The 99R offers a high tolerance level to radicicola phylloxera, but is very sensitive to gallicola phylloxera. It has a high resistance to downy mildew and good resistance to *Meloidogyne hapla* but only average resistance to *Meloidogyne incognita* and *Meloidogyne arenaria* nematodes. It is also sensitive to anthracnose. 99 R resists up to 25% total limestone, 14% active limestone, and a CPI of 20. It has moderate to strong resistance to drought, and it has shown to be sensitive to soil acidity and excess chlorides (INRA 2016). 99R cannot tolerate salt (Pongrácz 1983). It presents a good graft affinity. It has strong vigor and tends to delay the plant cycle of the grafts, and sometimes promotes coulure (shatter), which causes a failure to develop berries after flowering. Also, reported an increase in the risk of degeneration with 'Syrah'. 99 R internodes are of average length, with an average to large in diameter. It is a moderate wood producer and propagation by cuttings and grafting (INRA 2016).

(c) **110R (110 Richter, 110R)** (*V. berlandieri* 'Rességuier 2' × *V. rupestris* Martin)

The 110R rootstock presents good resistance to radicicola phylloxera (INRA 2016) but is very sensitive to gallicola phylloxera. Few root nodosities were observed in 110R potted vines (Whiting 1992). It has average resistance to the nematodes *Meloidogyne incognita* and *Meloidogyne arenaria*. It also has a high resistance to downy mildew and is not very affected by anthracnose. It also shows some resistance to *Phytophthora cinnamomi*. 110R is moderately adapted to limestone, and it has variable resistance to ferric chlorosis (depends on grafted variety). It is considered to be able to resist up to 17% active limestone and a CPI of 30. However, the threshold is really 5 to 7% when there is grafting with Syrah (and to a lesser extent with Viognier). 110R is very well adapted to drought but is very sensitive to excess humidity. 110R shows strong vigor (INRA 2016) but does not widely grow at present (Galet 2004). It tends to induce a delay in the plant cycle and maturation. It can sometimes favor the onset of color, particularly with 'Ugni blanc'. It shows very good affinity. It shows chlorosis for Syrah. Graft in-compatibility was also reported with Pinot noir. 110R has a moderate length of internodes with a quite large diameter. It is a low to moderate wood producer with occasionally a proportion of dry wood and needs care for the proper lignifying (INRA 2016). Vines start slowly but outgrow those on 99R or 101-14Mgt by the end of the first season (Pongrácz 1983). In Australia, 110R displays vein necrosis with low vigor and low pH juice (Walker 1991). It has low to moderate cutting and grafting capacity and needs longer stratification and more hormone (INRA 2016).

(d)   **140 Ru (140 Ruggeri)** (*V. berlandieri* 'Rességuier 2' × *V. rupestris* 'du Lot')

The 140Ru has good resistance to radicicola phylloxera and is sensitive to gallicola phylloxera. This rootstock shows high resistance to the nematode *Meloidogyne arenaria* and moderate resistance to nematode *Meloidogyne incognita*. It is also highly resistant to downy mildew and not sensitive to anthracnose. 140Ru presents high drought resistance (INRA 2016) but is susceptible to wet feet and not recommended for fertile sites (Walker, 1991). It has good adaptation to limestone soils. It resists up to 50% total limestone, 20% active limestone, and a CPI of 90. 140Ru well absorbs magnesium. 140Ru is a very high vigor and tends to delay the growing cycle. 140Ru has moderate-length internodes with a quite large diameter. This is a low to medium wood producer and shows low cutting results, which needs better stratification and using hormones (INRA 2016). It has very deep roots and is well-branched (Walker 1991). It is presenting moderate grafting results with some cultivars since the grafting union is weak (INRA 2016).

(e)   **775 P (775 Paulsen)** (*V. berlandieri* 'Rességuier 2' × *V. rupestris* 'du Lot')

Cosmo et al. (1958) considered it as a good rootstock. Pongrácz (1983) mentioned that the cuttings root and graft well and have a very good affinity with commercially cultivated *V. vinifera* varieties. In the fertile soil, the grafted vines on 775P present too vigorous, with a poor fruit set. It can adapt well to dry and compact soils during summer and can tolerate water shortage in the soil. It can support up to 17% active lime in the soil (Pàstena 1976). It shows a high yield under relatively saline soil (Southey and Jooste 1991).

(f)   **779 P (779 Paulsen)** (*V. berlandieri* 'Rességuier 2' × *V. rupestris* 'du Lot')

It has excellent induced vigor, phylloxera, and drought tolerance. It has a variable affinity with *Vitis vinifera* cultivars (VCR 2019). 'Tempranillo' and 'Cannonao' cultivars can show problems of affinity if grafted onto this rootstock (VCR 2007).

(g)   **1103P (1103 Paulsen)** (*V. berlandieri* 'Rességuier 2' × *V. rupestris* 'du Lot')

1103P presents a high resistance to radicicola phylloxera and is moderately suscep- tible to gallicola phylloxera. It has moderate resistance to the nematode *Meloidogyne incognita*, is sensitive to the *Meloidogyne arenaria*, and is highly resistant to downy mildew (INRA 2016). It is moderately tolerant of salt (Walker 1991) and absorbs magnesium well. 1103P resists up to 30% total limestone, 17% active limestone, and a CPI of 30. It is well adapted to drought and compact soil and the possible presence of substantial temporary spring humidity. In addition, it is well suited to acidic soil with a fairly good tolerance to chlorides. 1103P has considerable vigor and tends to produce sucker growth. It is a low to medium wood producer, and the quality of the cuttings is poor(twisted or broken wood). The propagation by cuttings capacity is moderate, and the grafting capacity is very good. Although, some graft affinity problems were reported with 'Tempranillo' (INRA 2016). Late maturity is induced in scions grafted to 1103P and is not recommended for the fertile sites (Walker 1991).

(h)   **1447P (1447 Paulsen)** (*V. berlandieri* × *V. rupestris* 'Martin')

The 1447P rootstock presents a high tolerance to radicicola phylloxera, and is suscep- tible to gallicola phylloxera. It has very good resistance to *Meloidogyne incognita* and *Meloidogyne arenaria* nematodes and shows high resistance to downy mildew. 1447P is moderate to well adapted to limestone soils. It resists up to 17% of total limestone. It is also resistant to drought. 1447P produces a high yield and displays strong vigor but has a somewhat slow plant growth rate. This rootstock has good propagation cuttings and grafting capacity (INRA 2016).

(i)   **Georgikon 121** (*V. berlandieri* × *V. rupestris*)

Georgikon121 rootstock shows excellent resistance to phylloxera, compared with 5C (Tóth and Kocsis 2003).

11.3.11   *V. berlandieri* × *V. rupestris* × *V. aestivalis* × *V. vinifera*

(a)   **US 8-7 (USVIT 8-7)** (Richter 99 (*V. berlandieri* Las Sorres × *V. rupestris* 'du Lot') × Jacquez (*V. aestivalis* × *V. vinifera*))

US 8-7 is a bred rootstock developed at the University of Stellenbosch, South Africa. A number of crossings were planted in the field trials, and US 8-7 is the only one that has been promised and commercialized on a larger scale (Teubes 2014).

11.3.12   *V. berlandieri* × *V. vinifera*

(a)   **BC2 (BC2, Berlandieri Colombard2, Colombard2)** (*V. berlandieri* × *V. vinifera* 'Ugni blanc B')

The BC2 rootstock has a moderate to strong tolerance against radicicola phylloxera, but is susceptible to downy mildew. It has good adaptation to limestone soil and resists up to 40% active limestone. Grape varieties grafted with this rootstock show a good speed of development. It has shown a moderate cutting and a good grafting capacity (INRA 2016).

11.3.13   *V. berlandieri* × *V. vinifera* × *V. berlandieri* × *V. longii*

(a)   **Fercal** (Berlandieri-Colombard 1 B (*V. berlandieri* × *V. vinifera* 'Ugni blanc B') × 31 Richter (*V. berlandieri* 'Rességuier 2' × *V. longii* Novo-mexicana)

Fercal has a very high tolerance to radicicola phylloxera and moderate sensitivity to gallicola phylloxera. It also presents a good tolerance to the nematodes *Meloidogyne arenaria* and *M. incognita* but has just moderate resistance to the *M. hapla*. Fercal is resistant to downy mildew and to anthracnose. It has very good resistance to chlorosis and its adaptation to limestone soils. It resists up to 60% total limestone, 40% of active limestone, and a CPI of 120. This rootstock is moderate to well resistant to drought and fairly well tolerates humid conditions in spring. Fercal shows moderate to strong vigor, and it has a moderate influence on the vegetative cycle and a good speed of growth and fruit set and produces good quality crops. This rootstock has moderate length and diameter of the internodes, and it is a moderate wood producer. It has very good cuttings and moderate grafting capacity. This rootstock shows a good graft affinity but has a problem absorbing magnesium from the soil, and the grafted vine can show deficiency especially if applied excessive potassium fertilizers (INRA 2016).

11.3.14   *V. berlandieri* × *V. vinifera* × *V. rupestris*

(a)   **1045P (1045 Paulsen)** (*V. berlandieri* 'Rességuier 2' × A × R#1 (*V. vinifera* Aramon × *V. rupestris* Ganzin)

1045P displays good adaptation to the relatively saline soils. 'Colombard' on 1045P had high pH and above-average total soluble solid concentration of berries and showed the highest wood producing compared with other rootstocks (1l03P, 775P, 99 R, 1l0R, 140 Ru, 143-B Mgt, 101-14 Mgt, Ramsey, 13-5 EVEX, Constantia

Metallica, Jacquez, Grézot 1, 216-3 Cl and US hybrids, 1-6, 3-6, 4-4, 2-1, 24-41, 24-10, 35-1-15, 16-13-23, 16-13-26 and 12-6-8) (Southey and Jooste 1991). 'Syrah' on 1045P displayed a high color intensity, polymerized pigment index, and total titratable acidity content (Dias et al. 2017).

11.4   *Vitis candicans*

11.4.1   *V. candicans*

(a)   **Leverkuhn** (*V. candicans* open pollinated)

It has shown a slight symptom of anthracnose (Mortensen 1980), but it is a source of resistance to Pierce's disease (Mortensen et al. 1977).

11.4.2   *V. candicans* × *V. rupestris*

(a)   **La Pryor** (Probably a hybrid of *V. candicans* and *V. rupestris*)

Created in Texas A & M University (Reynolds 2015), and while it is not entirely immune to the root rot organisms, it is promising rootstock for use in infested areas (USDA 1943).

11.4.3   *V. candicans* × *V. rupestris* × *V. riparia* × *V. rupestris*

(a)   **RS-2** (Ramsey (*V.* candicans × *V. rupestris*) × Schwarzmann (*V. riparia* × *V. rupestris*))

It is reported that RS-2 is only suppressed reproduction by the two resistance-breaking populations of *Meloidogyne arenaria* and can exhibit a more durable root-knot resistance than commercially available rootstocks (Anwar et al. 2002).

(b)   **RS-3** (Ramsey (*V. candicans* × *V. rupestris*) × Schwarzmann (*V. riparia* × *V. rupestris*))

RS-3 presents a broad-spectrum resistance to nematodes (FPS Grapes 2018). RS-3 is slightly susceptible to the citrus nematode, *Tylenchulus semipenetrans,* and it has slightly less resistance to ring nematode than RS-2, but it shows resistance to *Xiphenema index* and *Pratylenchus vulnus* (JUSTIA 2003a). It has a moderate vigor level and is suitable for coarse to fine sandy loam soils (FPS Grapes 2018). Strong and long lateral shoots are present under high vigor conditions (JUSTIA 2003a).

(c)   **RS-9** (Ramsey (*V. candicans* × *V. rupestris*) × Schwarzmann (*V. riparia* × *V. rupestris*))

It is considered a rootstock that has a wide resistance to nematodes. It shows resistance to gall-initiating *Meloidogyne arenaria* and non-galling *M. chitwoodi,* which attack 'Harmony'. It displays good resistance to *Xiphinema index,* and *Pratylenchus vulnus,* and slight susceptibility to the citrus nematode. The RS-9 shows wider resistance to nematode than VR O39-16, Freedom, Harmony, Ramsey, and Teleki 5C. Its vigor is moderate and is comparable to Schwarzmann rootstock. RS-9 has been suggested for dense planting, coastal valleys, and coarse-textured soils (JUSTIA 2003b).

11.5   *Vitis caribaea*

11.5.1   V. caribaea × V. riparia × V. rupestris

(a)   **IAC 572 (Jales)** (*V. caribaea* × 101–14 Mgt (*V. riparia* × *V. rupestris*))

This is created in Brazil. It is resistant to phylloxera, and nematodes, downy mildew, and fusariosis and has low resistance to anthracnose (Embrapa 2014b). IAC572 may have mealybug resistance capacity on mealybug (*Dysmicoccus brevipes*) (FPS Grapes 2018). It is adapted to clay, sandy, and acidic soils (Embrapa 2014b). It is vigorous rootstock (IAC 2019), but excessive vigor is the most limiting factor and requires special management practices to control vegetative growth (Dalbó and Feldberg 2016). It is easy to root (Embrapa 2014b). It is distinguished as a good rootstock for 'Italy', 'Ruby', 'Benitaka', 'Red Globe', and 'Centennial Seedless' grapevine cultivars in Brazil (IAC 2019).

11.5.1.1   *V. caribaea × V. vinifera*

(a)   **IAC 571-6 (Jundiaí)** (*V. caribaea* × *V. vinifera* Pirovano 57 (Bicane × Poete Matabon)')

IAC571-6 leaves are resistant to the main fungal diseases. It has some tolerance to the ground pearl (*Eurhizococcus brasiliensis*), which is a common pest in Brazil. It is vigorous, with good growth in both clay and sandy soils (Pommer 2000). This rootstock is a hybrid, which has a tropical species (*V. caribaea*) in his pedigree with no dormancy period and could be induced to early budding and bud break of the grafted vines (Dalbó and Feldberg 2016). The cuttings show an excellent setting and show a very good development on grafted union. It is recommended as the best rootstocks for 'Niagara Rosada' cultivar in Brazil (Pommer 2000).

11.6   *Vitis × champinii*

11.6.1   *V. × champinii*

(a)   **Dog Ridge (Dogridge)** (*V. × champinii*)

Dog Ridge is moderately phylloxera-resistant and has strong resistance to root-knot nematodes (UC Davis 2016). It has the same characters as Ramsey but is harder to manage (Walker 1991). Dog Ridge induces very high vigor in scions grafted on it, especially on deep, fertile soils. Therefore, because of its very high vigor and dense canopies, it induces rarely used now and is just suggested for infertile coarse-textured soils. This is difficult to propagate and graft (UC Davis 2016).

(b)   **Ramsey (Salt Creek)** (*V. × champinii*)

Salt Creek is a synonym for Ramsey. It is reported that the true Salt Creek is a selection of *Vitis doaniana* and is not used as a rootstock (UC Davis 2016). Ramsey has high resistance to phylloxera. However, the resistance may not be durable and could be rated differently. It is highly resistant to root-knot nematodes but has low resistance to dagger nematodes (Walker 1991). The coastal growers should consider other rootstocks for general nematode control (UC Davis 2016). It presents a moderate lime tolerance. It is well suggested for coarse-textured low fertility soils and recommended for sandy soils. Ramsey is susceptible to Zinc deficiency and has shown a problem with potassium absorb in some varieties (Walker 1991). Ramsey induces very high vigor in scions but slightly less than Dog Ridge (UC Davis 2016). It is mentioned that Ramsey is not vigorous in cooler areas (Walker 1991). Ramsey has a poor rooting rate (Walker 1991).

11.6.2   *V. × champinii × V. aestivalis × V. labrusca*

(a)   **Florilush** (Dog Ridge (*V. × champinii*) × Tampa (*V. aestivalis × V. labrusca*))

Florilush is an interspecific hybrid bred by the University of Florida as a bunch grape. It is resistant to nematodes and Pierce's disease (FPS Grapes 2018). It has a higher tolerance to grape root borer than 'Tampa'. It is recommended for Florida's deep, sandy soils and hot, humid climate. It also enhances vigor in young scions. 'Florilush' has a higher percentage of successful grafts and a lower tendency to produce sprouts below the graft union than 'Dog Ridge' (Mortensen et al. 1994).

11.6.3   V. × champinii × V. riparia × V. × champinii × V. berlandieri

(a)   **GRN-5 (9407-14)** (*V. × champinii* 'Ramsey' × *V. riparia 'Riparia Gloire'*) × *V. × champinii* c9021 (a form of *V. × champinii* that appears to integrate with *V. berlandieri*).

The GRN-5 has the highest level of nodosity-based phylloxera on roots. It has strong and wide nematode resistance. It has excellent root-knot and dagger nematode resistance. It can resist citrus nematodes and lesion nematodes and low numbers of ring nematodes. *V. riparia* was used as a parent of this rootstock to increase the rooting and grafting results. Cuttings produce medium-sized roots with relatively deep rooting angles (FPS Grapes 2018).

11.6.4   V. × champinii × V. labrusca × V. vinifera × V. labrusca

(a)   **Champanel** (*V. × champinii* × *V. labrusca* 'Worden' (*V. labrusca* 'Concord' O.P. (Catawba (*V. labrusca* × *V. vinifera* 'Semillon') × *V. labrusca*) × *V. labrusca* 'Cassady'))

Champanel displays very susceptible to root-knot nematodes but has a high consistent resistance to the disease. (Kamas 2014). It is resistant to Pierce's Disease, Black Rot, Downy Mildew, Anthracnose, and Powdery Mildew (McEachern 1997). The vine is extremely vigorous (McEachern 1997), but it presents a lower vigor than Dogridge, (Kamas 2014). Champanel grows well in a broad range of soils but has an iron chlorosis problem on high pH poorly drained soils (McEachern 1997).

11.6.5   V. × champinii × V. riparia

(a)   **K51-32** (V. × champinii × *V. riparia*)

The K51-32 is resistant to *Meloidogyne javanica* (Clingeleffer and Smith 2011) and root-lesion nematode (*Partylenchus vulnus*). It is sensitive to ring nematode (*Mesocriconema xenoplax*) and pin nematode (*Paratylenchus hamatus*) (Ferris et al. 2012). It has high potassium uptake and is moderately sensitive to salinity and sensitive to drought. It has similar vigor to Ramsey but is more easily propagated (Dry 2008).

(b)   **K51-40** (*V.* × *champinii* 'Dog Ridge' × *V. riparia* 'Gloire')

The K51-40 is resistant to *Meloidogyne javanica* (Clingeleffer and Smith 2011). It is sensitive to salinity (Dry 2008). Grafted 'Shiraz' on K51-40 shows low yield under non irrigated conditions (McCarthy 2008).

11.6.6   *V.* × *champinii* × *V. rufotomentosa* × *V. riparia*

(a)   **Kingfisher** (4-12A (*V.* × *champinii* Dog Ridge × *V. rufontomentosa*) × *V. riparia* 'Gloire')

Kingfisher was released from the USDA-ARS breeding program in 2010 (FPS Grapes 2018). It is an improved root-knot nematode-resistant rootstock (*Meloidogyne incognita*) (Owens 2012; FPS Grapes 2018). Kingfisher was recognized as a seedling due to its complete suppression of root-knot nematode reproduction in greenhouse evaluation (UC Davis 2016). Its root-knot nematode resistance was also confirmed in replicated tests of cutting grown plants. The nematode population used to confirm resistance was an N-virulent nematode population capable of feeding on and damaging N-allele grapevine rootstocks, such as Harmony and Freedom (UC Davis 2016). Kingfisher is easily rooted (100% success rate) from dormant cuttings and bench grafted to *V. vinifera* scions (UC Davis 2016).

11.6.7   *V.* × *champinii* × *V. rupestris*

(a)   **J 17-48 (J17-48)** (*V.* × *champinii* × *V. rupestris*)

It was bred by the University of California (Cirami 1994). Resistant to *Meloidogyne javanica* (Clingeleffer and Smith 2011). It shows no differences in yield between using Ramsey, Freedom, and J17-48 in Australia. It shows low juice pH and potassium (Cirami 1994).

(b)   **J 17-69 (J17-69)** (*V.* × *champinii* × *V. rupestris*)

It was bred by the University of California (Cirami 1994). J17-69 shows a recognizable ability to adequately resist root-knot nematodes. It shows no differences in yield with using Ramsey, Freedom, and J17-69 in Australia. It shows low juice pH and potassium combined (Cirami 1994).

11.7   *Vitis cinerea*

11.7.1   *V. cinerea*

(a)   **Cinerea 'Arnold'**

Cinerea 'Arnold' is a wild *Vitis cinerea*, which was considered as a grape rootstock. Carl Börner bred Cinerea 'Arnold' and crossed it with *Vitis riparia* in Germany. Helmut Becker later selected from those seedlings a completely phylloxera-resistant rootstock 'Börner'. Dry (2005) mentioned the bred rootstocks with phylloxera resistance based on the hypersensitive response elicited by *V. cinerea* 'Arnold'. Pavloušek (2012) also described a method for screening phylloxera resistance in rootstock hybrids by using *V. cinerea* 'Arnold'.

11.7.2   *V. cinerea* × *V. cinerea*

(a)   **M6262 (Merbein 6262)** (*V. cinerea* 'B58' × *V. cinerea* 'B194-1')

M6262 was bred in Australia and shows moderate tolerance to G4 and G30 strains of phylloxera, which indicates a potential of radicicola phylloxera developing, 'Shiraz' grafted on M6262 which was planted on an infested soil, showed no symptoms of nematode problem over 20 years. M6262 showed low-level reproduction of root-knot nematode for four pathotypes of *Meloidogyne javanica* and one of *M. incognita* in greenhouse trials. This is a low vigor rootstock. A trial in Australia showed that the pruning wood weights and crop yield of 'Shiraz' grafted to M6262 were lower than using Ramsey or 1103P. The M6262 is not recommended for saline situations. 'Shiraz' on M6262 accumulated significantly less potassium in berries leading to lower juice pH and thus less need for acid adjustments in winemaking, and presented lower color hue, higher color density, and higher total phenolics compared with using 1103P and Ramsey rootstocks (Walker et al. 2014).

11.8   *Vitis cordifolia*

11.8.1   *V. cordifolia* × *V. rupestris*

(a)   **Grasset 1** (*V. cordifolia* × *V. rupestris*)

It is recognized as a very vigorous rootstock (Viala and Rafaz 1903).

11.9   *Vitis labrusca*

11.9.1   *V. labrusca* × *V. riparia*

(a)   **Sonona** (*V. labrusca* 'Lady' × *V. riparia*)

It was created in South Dakota. Sonona is shown to control vigor and induce cold hardiness (Pool et al. 2005). The observed temperature of the low-temperature exotherm for the Sonona was −31.2 °C (Pool et al. 1990). Sonona shows very low drought tolerance (Pool 2004). The effect of shoot thinning of four commercial rootstocks (Sonona, 3309C, SO4, and 44-53M) on the production of wood suitable for grafting showed that the Sonona provided the largest quantity of graft-able wood per vine and could be grown without shoot thinning to reduce labor costs (Stobbs and Van Schagen 1987).

(b)   **Shakoka** (*V. labrusca* 'Lady' × *V. riparia*)

It is obtained from South Dakota. Shakoka is sensitive to phylloxera (Pool 2004). The observed temperature of the low-temperature exotherm (LTE) for the Shakoka was −28.5 °C (Pool et al. 1990). It is reported that 'Chardonnay' grafted on Shakoka presented a small canopy (Pool et al. 2005).

11.10   *Vitis lincecumii*

11.10.1   *V. lincecumii* × *V. rupestris* × *V. vinifera* × *V. simpsonii*

(a)   **Demko 10-17A (10-17A)** (America (open pollinated Jaeger 43 (*V. lincecumii* × *V. rupestris*) × Malaga (*V. vinifera*)) × *V. simpsonii*).

The United States Department of Agriculture, Agricultural Research Service (USDA-ARS) released this rootstock in 2012. Demko10-17A shows a very good nematode resistance. It has been tested and found to be free of known viruses (FPS Grapes 2018; Reynolds 2015).

11.11   *Vitis longii*

11.11.1   V. longii × V. riparia × V. rupestris

(a)   **216-3 Cl (216-3 Castel)** (1616 Couderc (*V. longii × V. riparia*) × *V. rupestris* 'du Lot')

The 216-3Cl rootstock shows a high tolerance to radicicola phylloxera, but shows only good tolerance to gallicola phylloxera. It is susceptible to *Meloidogyne arenaria* and *Meloidogyne incognita* nematodes. It appears to have a high resistance to downy mildew. 216-3Cl resists only up to 10% active limestone. It is fairly well adapted to drought and displays a low tolerance to chlorides. 216-3Cl is a strong vigor rootstock but is a low wood producer. 216-3Cl shows moderate cuttings and low grafting capacity (INRA 2016).

11.12   *Vitis monticola*

11.12.1   V. monticola × V. riparia

(a)   **188-04 Cl (188-04 Castel)** (*V. monticola × V. riparia*)

188-04Cl is a resistant rootstock to *Meloidogyne javanica* (Clingeleffer and Smith 2011).

(b)   **188-15 Cl (188-15 Castel, Castel 18,815, 18-815 Couderc, 18-815 Mgt)** (*V. monticola × V. riparia*)

Fingerprints of 188-15 Castel and 18-815 Couderc are matching (VIVC 2019). The 18-815C shows good phylloxera and lime resistance (Galet 2004). It shows a cold tolerance, and the low-temperature exotherm (LTE) for the 18-815C is −23.7 °C (Pool et al. 1990). The 188-15Cl has hermaphrodite flowers and makes the berries. It shows the highest total content of polyphenols compounds in the berries between 48 hybrid grapes (Liang et al. 2013). It is not reported if it can induce a higher level of polyphenolic content the scion or not.

11.13 *Vitis mustangensis*

11.13.1 *V. mustangensis × V. rupestris*

(a) **3-1A** (*V. mustangensis × V. rupestris*)

It is an interspecific *Vitis* hybrid selection with demonstrated resistance to N-virulent nematodes. It was used as a pollen parent for Matador and Minotaur rootstocks (Cousins 2012).

11.14 *Vitis riparia*

11.14.1 *V. riparia*

(a) **Riparia '1 G' (Riparia 1 Geisenheim)** (*V. riparia*)

This rootstock variety was selected out of 200 *Vitis riparia* seedlings in Geisenheim, Germany. The growing conditions of this variety are similar to Riparia Gloire but have smaller leaves (FPS Grapes 2018).

(b) **Riparia 'Crosby'** (*V. riparia*)

It has been reported that its rachis and pedicel are susceptible to powdery mildew (*Uncinula necator*) (Gee et al. 2008).

(c) **Riparia 'Gloire de Montpellier' (RGM, Portalis)** (*V. riparia*)

RGM is very highly tolerant to radicicola phylloxera and moderately well tolerant to gallicola phylloxera. It also has a quite tolerance to *Meloidogyne hapla* and moderate resistance to *Meloidogyne incognita* and *Meloidogyne arenaria* nematodes. RGM has a certain degree of tolerance to *Agrobacterium vitis*, and resists well downy mildew. RGM is adapted to acidic or decalcified soils and is very sensitive to chlorosis. It resists less than 15% total limestone, 6% active limestone, and a CPI of 5. It is well adapted to humidity but sensitive to drought (INRA 2016). It prefers fresh, deep, fertile soil well supplied with water (Pongrácz 1983). It is poorly adapted to clay and compact soils. It can easily absorb potassium and poorly absorb magnesium (show magnesium symptoms deficiency and black stem). RGM is a short-cycle rootstock (Galet 2004) and provides early ripening (Walker 1991). The grafted scion on RGM generally presents limited yield and shows very low vigor (INRA 2016) similar to or less than 101-14Mgt vigor, and it is used where scions are low in vigor (Perry and Sabbatini 2015). This rootstock has a high affinity for some cultivars such as 'Cabernet Sauvignon', 'Chardonnay', 'Merlot' and 'Sauvignon'. RGM is a good wood producer with long internodes, an average diameter, and has good cuttings and grafting capacity (INRA 2016). It shows good rooting and has a shallow fibrous root system (Walker 1991). Its trunk remains limited and shows differences in diameter with the grafts (INRA 2016).

(d)   **Riparia 'Grande Glabre'** (*V. riparia*)

This rootstock was created in France. It has some synonyms such as Golaya, Riparia 13, Riparia Arnaud, Riparia Grain Glabre, and Riparia Velikolisna Glatka (VIVC 2019). It is resistant to phylloxera, and very vigorous. The canes and internodes are long with a medium diameter (Viala and Ravaz 1903). 'Zinfandel' showed a fair yield when grafted on Grand Glabre but showed higher acidity and lower °Brix compared with RGM (Hayne 1896). It was the seed parent for some common rootstocks, such as 44-53M and 4010Cl (VIVC 2019).

(e)   **Riparia 'Montreal'** (*V. riparia*)

It is a wild selection from Quebec, Canada. It has a shorter vegetative cycle than Riparia Gloire de Montpellier (Pool et al. 2005).

(f)   **Riparia 'Scribner'** (*V. riparia*)

It is resistant to phylloxera and very vigorous with large leaves. It is superior in compact soil (Viala and Ravaz 1903).

(g)   **Riparia 'Simcoe'** (*V. riparia*)

Twelve rootstocks, including Simcoe 034, 089, 120, 123, 164, 494, 517, 634, 659, 732, 801, and 802 have been selected from nine hundred *Vitis riparia* accessions collected from southwest Ontario, Canada. They are under further evaluation for drought and cold tolerance (Rahemi et al. 2016) and have been disclosed by the University of Guelph in 2019. It has tested d negative for several viruses (RB, FL, PG, LR1,2,3,4) (Fisher 2021).

11.14.2   *V. riparia* × *V. cinerea*

(a)   **Börner (Borner)** (*V. riparia* '183 Geisenheim' × *V. cinerea* 'Arnold')

Börner is a rootstock created based on this idea to combine the positive viticultural characteristics of *Vitis riparia* with the complete phylloxera resistance potential of *Vitis cinerea* (JUSTIA 1994). Börner is introduced as one of the resistant grape rootstocks to phylloxera. This rootstock reacts with a hypersensitive reaction (HR), leading to local necrosis on leaves and roots and preventing the feeding and breeding of the insect (Börner 1943; Niklowitz 1955; Anders 1958). In Germany, Börner appears best on well-structured, stony, and dry soil types with low pH, especially on steep-sloped hills and terraces. In practice, it is one of the best presents on sites that suffer from soil exhaustion. It has good drought tolerance but shows low-medium chlorosis resistance. It can tolerate about 12% active lime. It is a medium-strong growth vigor rootstock with medium to good rooting capacity. Börner displays early to medium vegetation cycle and very good grafting affinity (Geisenheim 2016).

(b)   **Rici** (*V. riparia* 183 Geisenheim × *V. cinerea* 'Arnold')

It performs a medium to high drought resistance and a low to medium chlorosis resistance (Pavloušek 2012).

11.14.3   *V. riparia* × *V. cordifolia* × *V. rupestris*

(a)   **44-53 Malégue (44-53M)** (*V. riparia* Grand Glabre × 144 Malègue (*V. cordifolia* × *V. rupestris*)

44-53M has good tolerance to radicicola phylloxera (Galet 2004; Pongrácz 1983) and is not very susceptible to gallicola phylloxera. It performs moderate resistance to *Meloidogyne arenaria* nematodes, and it is susceptible to *Meloidogyne incognita* nematodes. It shows very good resistance to downy mildew (INRA 2016). It has good drought tolerance (Pongrácz 1983; Galet 2004; Drought 2016). The tolerance level of 44-53M to chlorosis is low to moderate. It is well adapted to fairly deep hard limestone soils, with a low tendency to induce chlorosis. It resists just up to 15% total limestone, 10% active limestone, and a CPI of 10 (INRA 2016). It is sensitive to magnesium deficiency (Galet 2004), and shows deficiency symptoms (INRA 2016). 44-53M is low to moderate rootstock vigor with balanced vegetative growth (INRA 2016). It is less vigorous than other *V. berlandieri* × *V. rupestris* hybrids such as 110R and 99R (Pongrácz 1983). 44-53M performs a good affinity to grafts and induces good quality crops. It is a moderate wood producer and has good cuttings and grafting capacity (INRA 2016).

(b)   **Riparia do Traviú (Riparia de Traviú, 106-8 Mgt, Millardet et de Grasset 106-8)** (*V. riparia* × Grasset 1 (*V. cordifolia* × *V. rupestris*))

106-8 Mgt is used as a raisin grape and also as rootstock. It is resistant to phylloxera (Bioletti 1908). It is considered dry resistant and also adapted to wet and clay soils (Nogueira 1984). It is recommended for clay soils that become hard after rain and easily dry out. It is a vigorous rootstock (Bioletti 1908). It shows low rooting of herbaceous cuttings, even applying auxins (Villa et al. 2003).

11.14.4   *V. riparia* × *V. cordifolia* × *V. rupestris* × *V. berlandieri*

(a)   **M1** (106-8 Mgt [*V. riparia* × Grasset 1 (*V. cordifolia* × *V. rupestris*)] × *V. berlandieri* 'Rességuier 1')

M1 is highly resistant to iron chlorosis and has medium resistance to salt. It is a low vigor rootstock, which was created by the University of Milan, Italy (Bavaresco et al. 2015).

11.14.5 *V. riparia × V. cordifolia × V. rupestris × V. caribaea*

(a) **IAC 766 (Campinas)** (Riparia do Traviú [106–8 Mgt] (*V. riparia* × Grasset 1 (*V. cordifolia* × *V. rupestris*)) × *V. caribaea*)

The IAC766 has high resistance to radicicola phylloxera, nematodes, downy mildew, and fusariosis. It is medium resistant to anthracnose. It is adapted to clay, sandy, and acidic soils (Embrapa 2014c). Easy to root (Embrapa 2014c). IAC766 is a hybrid of tropical species, which have no dormancy period, and could be induced to the early budding of the grafted vines (Dalbó and Feldberg 2016). It is considered as a good rootstock for 'Italy', 'Ruby', 'Benitaka', 'Red Globe', 'Centennial Seedless', 'Patricia', 'Maria', 'Paulistinha', 'Niagara', and 'Maximo' in Brazil (IAC 2019).

11.14.6 *V. riparia × V. labrusca*

(a) **Beta** ((*V. riparia* 'Carver') × (*V. labrusca* 'Concord')

Beta was bred in the U.S. and used as a table grape and also rootstock. It is considered a cold-resistant rootstock in China (Shao-Hua 2001).

(b) **Clinton** (*V. riparia* × *V. labrusca*)

Clinton shows a good joining when grafted with most of the European grape varieties, but it is sensitive to calcareous soils. It shows a good rooting from cuttings (Viala and Ravaz 1903). It has been used as a parent for many hybrids, such as 1613C, Harmony, LN33, Othello, and Vialla (VIVC 2019).

(c) **Vialla** (open pollinated Clinton (*V. riparia* × *V. labrusca*))

Vialla shows a low to moderate tolerance to radicicola phylloxera. Therefore, it should be used under free phylloxera soils. It is highly resistant to gallicola phylloxera and presents good resistance to downy mildew. It is also moderately resistant to *Meloidogyne arenaria* nematodes. Vialla is extremely susceptible to chlorosis, and it resists only up to 4% active limestone and a CPI of 2. It is susceptible to drought and must be used in good irrigation conditions. This rootstock is well adapted to acid and decalcified soils. Vialla shows good compatibility and good affinity with grafts Vialla is fairly strong, and induces early vegetative cycles, and tends to produce low acidic crops. This rootstock is a low wood producer and presents a good cutting and grafting capacity (INRA 2016).

11.14.7   *V. riparia* × *V. rupestris*

These hybrids root and graft well but have moderate to poor tolerance to lime and low adaptation to drought. In general, they reduce vigor but are not well adapted to Mediterranean conditions, and their favor is deeply fertile soils (Walker 1991).

(a)   **101-14 Mgt (MG 101-14, 101-14 Millardet et de Grasset)** (*V. riparia* × *V. rupestris*)

The 101-14Mgt shows good tolerance to radicicola phylloxera and moderate tolerance to gallicola phylloxera. It is tolerant to nematodes *Meloidogyne incognita* and *Meloidogyne arenaria*. It has good resistance to downy mildew but is fairly sensitive to anthracnose. 101-14 It is well adapted to waterlogging and sensitive to drought (Walker 1991; Pongrácz 1983; INRA 2016). It has less drought tolerance than 3309C (Galet 2004). It is low tolerance to chlorosis, and it resists up to 20% total limestone, 9% of active limestone, and a CPI of 10. It is fairly sensitive to soil acidity along with copper toxicity and bore deficiency (INRA 2016). It is an admired rootstock for clay soils (Perry and Sabbatini 2015; INRA 2016). It can absorb magnesium well. It is considered as a limited vigor rootstock with moderate yield and usually shows very good affinity with the grafts, which induces the early plant cycle. 101-14Mgt is recommended as a rootstock for 'Cabernet franc', 'Cabernet Sauvignon', 'Chardonnay', 'Sauvignon', and 'Syrah'. 101-14Mgt has a good budding and grafting capacity (INRA 2016).

(b)   **3306C (3306 Couderc)** (*V. riparia* tomenteuse × *V. rupestris* Martin)

It has good phylloxera resistance (Clarke 2004). It is resistant to *Meloidogyne javanica* (Clingeleffer and Smith 2011). The 3306C is a good rootstock in soils with more than >75 cm depth with a uniform or gradational profile of sand, loam, or clay (Pavloušek 2013). It is reported that 3306C is less vigorous than 3309C. It is a low wood yield producer but has an easy root and graft capacity. More vines with dying arm symptoms were recorded in 'Chenin blanc' grafted on 3306C compared with ten other rootstocks (Ferreira 1985). The 3306C shows were advancing ripeness in New Zealand (Clarke 2004). When 'Fujiminori' was grafted on seven different rootstocks (3309 C, 3306 C, 101-14 Mgt, 5C, 8B, SO4, and 420A) presented the highest level of anthocyanin in berry skin on 3306C (Kubota et al. 1993).

(c)   **3309C (3309 Couderc)** (*V. riparia* tomenteuse × *V. rupestris* Martin),

The 3309C shows a good tolerance level to radicicola phylloxera but is sensitive to gallicola phylloxera. It is sensitive to *Meloidogyne arenaria* and *Meloidogyne incognita* nematodes. It has a definite tolerance to *Agrobacterium vitis*. It is also well resistant to downy mildew but sensitive to anthracnose. The 3309C appears to be well suited to acidic soil and has slow to moderate tolerance to chlorosis with a resistance up to 20% total limestone, 11% of active limestone, and a CPI of 10 (INRA 2016). It is adapted to well-drained soils (Pongrácz 1983) but sensitive to drought, especially during the growing cycle. The 3309C easily absorbs magnesium but has little more problem absorbing potassium. This rootstock is sensitive to salinity. 3309C is well adapted to deep soils. It shows low to moderate vigor rootstock (INRA 2016) and slightly more vigorous than those grafted on 101-14Mgt (Perry and Sabbatini 2015). It is a low to moderate wood producer, with excellent propagation capacity for cuttings and grafting (INRA 2016). Cane hardiness is very good on this rootstock (Howell 1987). The good quality crops observed for 'Cabernet franc', 'Chardonnay', 'Colombard', 'Merlot', 'Pinot noir', 'Riesling' 'Cabernet sauvignon', 'Sauvignon', and 'Syrah'. At the same time, some problems were reported with some cultivars such as 'Cabernet sauvignon', 'Syrah', and 'Sauvignon'.

(d)   **6736 Cl (6736 Castel)** (*V. riparia* × *V. rupestris* 'du Lot')

The 6736Cl rootstock is one of the nine most used rootstocks in Spain (Ortiz et al. 2012). It is similar to that shown 3309C, but with often larger and more armored leaves, and with the likelihood to a little bit more open petiolar sinus (Laguna Lumbreras 2004b).

(e)   **Schwarzmann** (*V. riparia* × *V. rupestris*)

It is highly resistant to phylloxera (Walker 1991; Perry and Sabbatini 2015) and is resistant to ectoparasitic nematodes; however, it has moderate levels of *Xiphinema index* feeding (Davis 2016). It is suggested for deep fertile soils (Davis 2016), has moderate lime tolerance, and has high potassium uptake. It should not be used if summer drought is common. It is suggested when the high acidity of juice is not a problem (Walker 1991). It can induce moderate vigor (Walker 1991; Davis 2016) and is slightly more vigorous than 101-14Mgt and 3309C (Perry and Sabbatini 2015). It is easy to propagate and has a good rooting capacity (Walker 1991).

11.14.8 *V. riparia* × *V. rupestris* × *V. mustangensis* × *V. rupestris*

(a)   **Matador** (101-14 Mgt (*V. riparia* × *V. rupestris*) × 3-1A (*V. mustangensis* × *V. rupestris*))

It is a seedling selection released from the USDA ARS breeding program. It is a root-knot nematode-resistant rootstock and shows a complete suppression of root-knot nematode reproduction in greenhouse evaluation. A N-virulent nematode population was used to evaluate resistance capable of feeding on and damaging N-allele grapevine rootstocks. Root-knot nematode resistance was confirmed in replicated tests of cutting grown plants. Matador is easily rooted from dormant cuttings and bench grafted to *Vitis vinifera* scions (FPS Grapes 2018). It shows a 73% success rate of rooting dormant cuttings. It shows a fruit to pruning weight ratio of 9.4 (Owens 2012).

(b)   **Minotaur** (101-14 Mgt (*V. riparia* × *V. rupestris*) × 3-1A (*V. mustangensis* × *V. rupestris*))

Minotaur is a full sibling rootstock with Matador, and they had the same seed and pollen parent (FPS Grapes 2018). It is also a root-knot nematode-resistant rootstock released from the USDA ARS. Minotaur is also easily rooted from dormant cuttings and bench grafted to *Vitis vinifera* scions (FPS Grapes 2018). Minotaur shows a 92% success rate of rooting dormant cuttings, and its fruit to pruning weight ratio is 8.8 (Owens 2012).

11.14.9   *V. riparia* × *V. rupestris* × *V. vinifera* × *V. vinifera*

(a)   **Kraus1 (LE-K/1, K-1)** ((101-14 Mgt (*V. riparia* × *V. rupestris*) × *V. vinifera* Ortlieber) × Svaetovavrinecke (*V. vinifera* 'Saint Laurent')

Kraus1 has moderate resistance to phylloxera. It is resistant to powdery mildew (*Uncinula necator*) and lower to downy mildew (*Plasmopara viticola*). It has drought tolerance and therefore should be used on sandy and gravel soils. It is not recommended for calcareous soils due to its tolerance of only up to approximately 7% active Ca in soil. It is very vigorous, and grafted varieties show moderately vigorous to vigorous. Affinity to *Vitis vinifera* varieties is good. It is suitable for fertile and large-grape varieties. It can induce the ripening of grapes and maturing of wood. It usually has good resistance to frost. It has a good cutting and rooting capacity (Šimek 2017).

11.14.10   *V. riparia* × *V. vinifera* × *V. rupestris*

(a)   **4010Cl (4010 Castel)** (*V. riparia* Grand Glabre × A × R1 (*V. vinifera* Aramon × *V. rupestris* Ganzin))

4010Cl has a moderate tolerance to root and gallicola phylloxera. Therefore can be used only under unfavorable conditions for the phylloxera. It has a good resistance to *Meloidogyne arenaria* and *Meloidogyne incognita* nematodes. It has moderate resistance to downy mildew. It has a low tolerance to chlorosis, and it resists just up to 6% active limestone. 4010Cl cannot absorb magnesium very well. It is recommended to acid soils, very little or not limestone with enough water source. 4010Cl is a strong vigor rootstock. This rootstock tends to favor fertility. 4010Cl is a moderate wood producer but displays good propagation by cuttings and grafting capacity (INRA 2016).

11.15 *Vitis rotundifolia*

11.15.1 *V. rotundifolia* × *V. rotundifolia*

(a)   **MS 27-14 (MS27-14)** (*V. rotundifolia* × *V. rotundifolia*) (Syn. *Muscadinia rotundifolia* × *M. rotundifolia*)

MS27-14 is known to have resistance to *Meloidogyne incognita, M. arenaria,* and *M. javanica.* In a survey on grapevine rootstocks resistant to the root-knot nematode (*M. javanica*), no egg masses were found on the MS27-14 (Smith et al. 2017).

11.16 *Vitis rufotomentosa*

11.16.1 *V. rufotomentosa* × *V.* × *champinii* × *V. riparia* × *V. monticola*

(a)   **GRN-3 (9365-43)** (*V. rufotomentosa* × (*V.* × *champinii* 'Dog Ridge' × *V. riparia* 'Riparia Gloire')) × *V. monticola* 'c9038' (a wild collection))

GRN-3 generally has excellent nematode resistance. However, it has moderate resistance to ring nematode. It is a moderate vigor rootstock. Its cuttings shows very well root capacity, and its rooting angles are intermediate in-depth (FPS Grapes 2018).

11.16.2 *V. rufotomentosa* × *V.* × *champinii* × *V. riparia* × *V. monticola*

(a)   **GRN-4 (9365-85)** (*V. rufotomentosa* × (*V.* × *champinii* 'Dog Ridge' × *V. riparia* 'Riparia Gloire')) × *V. monticola* 'c9038' (a wild collection))

GRN-4 is a sibling of GRN-3. It has good resistance to root-knot and dagger nematodes and also resists citrus and lesion nematodes, but it is susceptible to ring nematodes. GRN-4 has slightly lower nematode resistance than GRN-3 but is a more vigorous rootstock. It has well root capacitys, and rooting angles are intermediate in length (FPS Grapes 2018).

11.16.3   *V. rufotomentosa* × *V.* × *champinii* × *V. riparia* × *V. riparia*

(a)   **GRN-2 (9363-16)** (*V. rufotomentosa* × (*V.* × *champinii* 'Dog Ridge' × *V. riparia* 'Riparia Gloire')) × *V. riparia* 'Riparia Gloire'

GRN-2 is susceptible to citrus and ring nematodes but has excellent resistance to root-knot and dagger nematodes. Its roots and grafts easily. GRN-2 has long shoots and internodes and a few laterals. It produces roots with relatively shallow rooting angles (FPS Grapes 2018).

11.17   *Vitis rupestris*

11.17.1   *V. rupestris*

(a)   **Rupestris 'du Lot' (Du Lot, de Lot, Saint George, St. George)** (*V. rupestris*)

It is fairly tolerant to radicicola phylloxera (Walker 1991; INRA 2016) but not as strong as *V. riparia* (Whiting 1992). Considerable numbers of nodosities were reported on potted plants but not in the field (Whiting 1992). It is very sensitive to gallicola phylloxera (INRA 2016). Poor nematode resistance (Walker 1991), it is sensitive to *Meloidogyne arenaria* and *Meloidogyne incognita*, but somewhat tolerant to *Meloidogyne hapla, Phytophthora cinnamomi,* and *Agrobacterium vitis.* It is sensitive to anthracnose but shows high resistance to downy mildew (INRA 2016). Slightly tolerant of chlorides and resists up to 25% total limestone, 14% active limestone, and a CPI of 20. Rupestris du Lot is recommended for soils with very little to no limestone. Its resistance to drought is moderate (INRA 2016), especially on well-drained non-restricting soils (Walker 1991). Avoid planting in compact soil. It can absorb potassium fairly well. It is a high vigor rootstock with short internodes and low to moderate cane diameter. It is a low wood producer but is good for propagation by cuttings or grafting (INRA 2016) with excellent rooting (Walker 1991). It has a good affinity with grafts such as 'Grenache' and 'Ugni blanc' (INRA 2016). This rootstock induces moderate to poor yields with high pH and potassium juices with high pH (Walker 1991).

(b)   **Rupestris 'Metallique' (Rupestris Metallica, Constantia Metallica)** (*V. rupestris*)

It has resistance to the phylloxera and slight resistance to drought. It can tolerate more calcareous clay soils than Rupestris 'du Lot'. Its vigor is a little lower than Rupestris 'du Lot' (Viala and Ravaz 1903).

(c)   **Rupestris 'Ganzin'** (*V. rupestris*)

It is used as the pollen parent for many common rootstocks such as A × R#1, A × R#9, and 1202C (VIVC 2019).

(d)   **Rupestris 'Mission'** (*V. rupestris*)

It showed a fair growth with good yield when it was used as a rootstock for the 'Zinfandel' variety (Hayne 1896).

(e)   **Rupestris 'Martin'** (*V. rupestris*)

It is very vigorous, with long canes and their leaves larger than Rupestris 'du Lot'. Root easily, but the grafting is harder than Rupestris 'du Lot' (Bioletti 1908). It is used as the pollen parent for many common rootstocks such as 110 R, 1447 P, 3306 C, 3309 C, and 57R (VIVC 2019).

11.17.2   *V. rupestris* × *V. rupestris*

(a)   **Schabort 1 (Schabort I)** (*V. rupestris*)

It shows the sensitivity to *Phytophthora cinnamomi* (Marais 1983).

11.17.3   *V. rupestris* × *V. berlandieri*

(a)   **Vivet 15** (*V. rupestris* × *V. berlandieri*)

It shows a relative sensitivity to *Phytophthora cinnamomi* (Marais 1983).

11.17.4   *V. rupestris* × *V. monticola*

(a)   **160-19 C (160-19 Couderc)** (*V. rupestris* 'Fort Worth' × *V. monticola*)

Susceptible to root-knot nematode (*Meloidogyne javanica*), and has shown an ability to reproduce root-knot nematode eggs in a glasshouse assay (Clingeleffer and Smith 2011).

11.17.5   *V. rupestris* × *V. rotundifolia*

(a)   **GRN-1 (8909-05)** (*V. rupestris* 'A. de Serres' × *V. rotundifolia* 'Cowart')

This rootstock has sterile flowers. This is the most resistant rootstock between the five nematode-resistant rootstocks released by Walker (2009). It has extremely strong and broad nematode resistance, and it is one of the rare sources of resistance to ring nematode. It has moderately-long shoots with shorter internodes and more laterals than the other four rootstocks. It has 80% rooting and grafting success from dormant cuttings (FPS Grapes 2018).

11.18   *Vitis simpsonii*

11.18.1   *V. simpsonii* × *V. vinifera*

(a)   **Lake Emerald** (*V. simpsonii* 'Pixiola' × *V. vinifera* 'Golden Muscat')

Lake Emerald has been recommended as Pierce's disease-resistant rootstock for Florida (Mortensen and Stover 1982).

11.19   *Vitis solonis*

11.19.1   *V. solonis* × *V. cordifolia* × *V. rupestris*

(a)   **202-4 Mgt** (*V. solonis* × (*V. cordifolia* × *V. rupestris*))

It is adapted to clay soil and especially non-calcareous clay soils. It can resist humidity in the soil and adapts to dry, poor, and hot soils (Bioletti 1908).

11.19.2   *V. solonis* × *V. riparia*

(a)   **1616C (1616 Couderc)** (*V. solonis* (*V. riparia* × *V. longii*) × *V. riparia* Gloire de Montpellier)

The 1616C rootstock has high tolerance to radicicola phylloxera and good tolerance to gallicola phylloxera. It has good resistance to *Meloidogyne arenaria* and *Meloidogyne incognita* nematodes and shows good resistance to downy mildew and anthracnose. 1616C has low to medium tolerance to chlorosis, and it resists only 11% active limestone. It has a tolerance to chlorides and fairly good adaptation to humidity and to sandy clay soils. The vigor of 1616C is low (INRA 2016). It is reported that the Davis selection of 1616C is different from the French selection of 1616C, and it is more vigorous and acts much like a more vigorous Riparia Gloire (Walker 1991). 1616C is a moderate wood producer and has a good cutting and grafting capacity. This rootstock produces good-quality products (INRA 2016).

(b)   **1616E (Couderc 1616E)** (*V. solonis* × *V. riparia*)

1616E refers to a selection of 1616 Couderc made and used in Germany (Galet 2004; Pool et al. 2005). It has moderate to good radicicola phylloxera resistance (Pongrácz 1983; Galet 2004) and moderate lime tolerance (Pongrácz 1983; Galet 2004), up to 11% lime (Pongrácz 1983). It can induce early scion maturity and tolerates wet soils (Galet 2004; Pool et al. 2005) and poorly drained soils, but it is sensitive to drought. It can be used in fertile sandy and slightly saline soils (Pongrácz 1983; Pool et al. 2005).

(c)   **Sori** (*V. solonis* × *V. riparia* 1 Geisenheim)

The risk of botrytis infection is reduced when using the Sori as a rootstock. It has good chlorosis resistance, with tolerance up to 15% free active lime. Sori withstands ground humidity conditions, especially on heavy, clayey soils. Sori is medium drought tolerant but performs well under temporary drought conditions on deep soils. It has medium growth vigor, and it is weaker than SO4. Its vegetation cycle is fairly short, and its wood ripens soon. It has a good rooting capacity and shows good grafting affinity, especially to coulure (shatter) sensitive varieties. Sori induces smaller and looser berries (Geisenheim 2016).

11.19.3   *V. solonis* × *V. riparia* × *V. labrusca* × *V. vinifera*

(a)   **1613C (1613 Couderc)** (*V. solonis* × Othello (Clinton (*V. riparia* × *V. labrusca*) × Black Hamburg (*V. vinifera*))

It has low to moderate phylloxera resistance and moderate to high nematode resistance (Robinson and Harding 2016). It is resistant to *Meloidogyne javanica* nematode (Clingeleffer and Smith 2011). The 1613C prefers fertile, sandy, loam soils (Robinson and Harding 2016).

11.19.4   *V. solonis* × *V. riparia* × *V. labrusca* × *V. vinifera* × *V.* × *champinii*

(a)   **Harmony** (Open-pollinated seedling of 1613C × Open-pollinated seedling of Dog Ridge (*V.* × *champinii*) = 1613 C (*V. solonis* (*V. riparia* × *V. longii*) × Othello (Clinton (*V. riparia* × *V. labrusca*) × Black Hamburg (*V. vinifera*)) × Dogridge (*V.* × *champinii*)

It is not resistant to phylloxera (Pongrácz 1983; Davis 2016) and is especially not recommended in clay loam soils since it promotes phylloxera (Davis 2016). It has good nematode tolerance (Pongrácz 1983; Galet 2004; Davis 2016). Its resistance to root-knot nematodes is based on a single gene (Davis 2016). It is moderate (Walker 1991) to high vigor (Walker 1991; Davis 2016) but has lower vigor and yield than Freedom rootstock. It is designed for sandy soils (Walker 1991).

(b) **Freedom** (Open-pollinated seedling of 1613C × Open-pollinated seedling of Dog Ridge (*V.* × *champinii*))

It is susceptible to phylloxera (Perry and Sabbatini 2015) and is not recommended for clay loam soils that promote phylloxera (Davis 2016). It is an excellent nematode-resistant rootstock (Davis 2016). It has high root-knot and moderate dagger nematode resistance (Walker 1991). It has high resistance to Armillaria root rot (*Armillaria mellea*) (Perry and Sabbatini 2015). Freedom is very sensitive to virus-infected scion wood, which can cause graft incompatibility (Davis 2016). It is moderate (Walker 1991) to high vigor (Walker 1991; Perry and Sabbatini 2015). It shows high pH and potassium problems in high vigor areas and is suggested as a moderate vigor in cooler sites (Walker 1991). It is the same with Harmony, designed for sandy soils (Walker 1991). It is an excellent choice for coarse-textured soils with relatively low fertility (Davis 2016).

11.19.5 *V. solonis* × *V. riparia* × *V. labrusca* × *V. vinifera* × *V. vinifera*

(a) **LN33** (1613 Couderc (*V. solonis* × Othello (Clinton (*V. riparia* × *V. labrusca*) × Black Hamburg (*V. vinifera*)) × *V. vinifera* 'Sultana')

LN33 is resistant to *Meloidogyne javanica* (Clingeleffer and Smith 2011). It has been suggested as an index for rapidly diagnosing grapevine corky-bark virus-like disease by in Vitro Micrografting (Tanne et al. 1993).

It has also been used for early detection of grapevine leafroll virus in *Vitis vinifera* by using in Vitro micrografting method. When leafroll infected scion material was micrografted on vines-free rootstock, the rootstock leaf turned red on it showed leaf rolling within 2–3 weeks (Pathirana and McKenzi 2005).

11.20 Vitis vinifera

11.20.1 V. vinifera × V. amurensis × V. riparia × V. rupestris

(a)   **Amos** (Severnyi (*V. vinifera* 'Seyanets Malengra' × *V. amurensis*) × Schwarz-mann (*V. riparia* × *V. rupestris*)

Amos is tolerant to phylloxera. Tolerates the active calcium in the soil up to about 10%, is suitable for lighter, sandy, loamy, sandy, and gravel soils with the lower calcium content. In heavy and damp soils, other rootstocks give better results. The rootstock also withstands drought for a short time. The frost resistance of this root-stock variety is very good. Growing is medium to vigorous, and induces the yield of varieties, and shortens the ripening period. It is a good wood producer, and it makes high-yield cuttings and has good rooting. Its affinity to *Vitis vinifera* varieties is good (Šimek 2017).

11.20.2 V. vinifera × V. berlandieri

(a)   **41 B Mgt (41B, 41 B Millardet et de Grasset)** (*V. vinifera* 'Chasselas Blank' × *V. berlandieri*)

It has moderate to high tolerance to radicicola phylloxera and high tolerance to gallicola phylloxera. It is also sensitive to the *Meloidogyne incognita, Meloidogyne arenaria,* and *Meloidogyne hapla* nematodes, and *Agrobacterium vitis*. It is sensitive to downy mildew (INRA 2016) but shows the highest resistance to rust (*Phakopsora euvitis*) (Hennessy et al. 2007). It is resistant to chlorosis, has good adaptation to limestone soils, and resists up to 60% total limestone, 40% active limestone, and a CPI of 60 (INRA 2016). Therefore, it should be grown where the lime tolerance is needed (Pongrácz 1983). It can absorb magnesium well from the soil. It has moderate drought resistance and is sensitive to temporary excess humidity conditions in the spring. It is not very adapted to compact soils. It shows a moderate vigor inducing by 41B (Galet 2004; INRA 2016) to strong (INRA 2016). 41B presents a good affinity with varieties, but some problems have been reported with 'Merlot' and 'Pinot noir'. It tends to delay the vegetative cycle of scions. It promotes grape cluster compactness with low sugar and slightly more acidic crops. It is a low to moderate wood producer with moderate to high length and diameter of the internodes. The cuttings capacity is low to moderate, but it shows a good grafting capacity (INRA 2016).

(b)   **5A-MZ (5A Martinez-Zaporta)** (*V. vinifera* × *V. berlandieri*)

It is an offspring of 41B Mgt, which was obtained by self-fertilization (Laguna Lumbreras 2004b) and was introduced by Hidalgo and Candela (1979) in Spain.

(c)   **333 EM (333EM, 333 Ecole De Montpellier, Foex 333)** (*V. vinifera* 'Cabernet-Sauvignon' × *V. berlandieri*)

It has a moderate to high tolerance to radicicola phylloxera and high tolerance to gallicola phylloxera. It is sensitive to the nematodes *Meloidogyne incognita* and *Meloidogyne arenaria*. It displays a moderate resistance to downy mildew but is very sensitive to leaf blister mites. It has resistance to chlorosis and very good lime tolerance and is used primarily in places with very high lime content (Galet 2004) where the lime tolerance exceeds the phylloxera susceptibility (Pongrácz 1983). It resists up to 60% total limestone, 40% active limestone, and a CPI of 70. It is sensitive to chlorides. It is also well adapted to drought and shallow soils. It is adapted to temporary excess humidity in the spring. It is a low wood producer and shows moderate cutting and a high grafting capacity. It can produce fairly high yields, and under not very fertile or limiting conditions, it produces good quality products but sometimes induces the onset of coulure (shatter) (INRA 2016).

(d)   **Evex 13-5** (*V. vinifera* × *V. berlandieri* 'Rességuier 2')

This rootstock originated in Spain. It showed contamination by the Arabis mosaic virus (FPS Grapes 2018). Its leaves appearance resembles 5A-MZ but is wider (Laguna Lumbreras 2004b).

11.20.3   *V. vinifera* × *V. berlandieri* × *V. berlandieri*

(a)   **M4** (41B (*V. vinifera* × *V. berlandieri*) × *V. berlandieri* 'Rességuier 1')

It is a medium to high vigor, optimal resistance to drought, and high resistance to salt, which was created by the University of Milan, Italy (Bavaresco et al. 2015).

11.20.4   *V. vinifera* × *V. riparia*

(a)   **26 G (Geisenheim 26)** (*V. vinifera* 'Schiava Grossa' × *V. riparia*)

Developed in Geisenheim (Germany). Until after the second world war, it was one of the most important varieties, but because of its low phylloxera tolerance, this variety was eliminated in the following decades and replaced with the more tolerant rootstocks (Becker et al. 2010).

(b)   **143A Mgt (Millardet et Grasset 143 A, Aripa)** (*V. vinifera* 'Aramon' × *V. riparia*)

Developed in Geisenheim (Germany), and was also eliminated and no longer recommended due to some disadvantages (low phylloxera tolerance), and replaced with the more tolerant rootstocks (Becker et al. 2010).

(c)   **143B Mgt (Millardet et Grasset 143 B)** (*V. vinifera* 'Aramon' × *V. riparia*)

It is susceptible to phylloxera and nematodes and has good tolerance to salinity and drought. It is well adapted to heavy, wet, and pot-clay soils with poor drainage. It shows moderate graft compatibility and long-term affinity. It is rarely used for wine grapes but performs excellently with raisin grapes (van Schalkwyk 2017).

11.20.5   *V. vinifera* × *V. riparia* × *V. rupestris*

(a)   **Golia** (Castel 156-12 (*V. vinifera* 'Carignan' × *V. riparia*) × *V. rupestris* 'du Lot')

It shows high resistance to chlorosis and high to very high resistance to drought (Pavloušek 2013). Golia showed a good affinity with 'Lambrusco grasparossa' (*V. vinifera*) with a good productive response (Credi et al. 1991).

11.20.6   *V. vinifera* × *V. riparia* × *V. rupestris* × *V. cinerea*

(a)   **IAC 313 (Tropical)** (Golia (Castel 156-12 (*V. vinifera* 'Carignan' × *V. riparia*) × *V. rupestris* 'du Lot') × *V. cinerea*)

It has high resistance to radicicola phylloxera and nematodes (Embrapa 2014a), but it is susceptible to *Meloidogyne javanica* (Clingeleffer and Smith 2011). It is highly resistant to downy mildew and presents medium resistance to Anthracnose (Embrapa 2014a). It is adapted to clay, sandy, and acidic soils (Embrapa 2014a). It is an easy to root (Embrapa 2014a), and has a good affinity with several varieties such as 'Italy', 'Ruby', 'Patricia', 'Benitaka', 'Red Globe', 'Paulistinha', and 'IAC138-22 Máximo' (IAC 2019). Since IAC313 is a hybrid of tropical species, it has no dormancy period and could be induced to the early bud-break of the scion (Dalbó and Feldberg 2016). It has been used in some regions in Brazil as the main rootstock (IAC 2019).

11.20.7   *V. vinifera* × *V. rotundifolia*

(a)   **DRX 55** ((*V. vinifera* × *V. rotundifolia*) open-pollinated)

DRX55 shows complete resistance to root-knot nematode reproduction but doesn't have good resistance to chloride from lamina (Clingeleffer and Davis 2012). It has high drought resistance and presents a relatively high cold resistance (Alexandrov 2017).

(b) **VR O39-16 (039-16, OLMO VR O39-16)** (*V. vinifera* 'Almeria' × *V. rotundifolia* #1)

O39-16 has proven to be resistant to phylloxera in the lab and field. It is a hybrid rootstock that developed as a rootstock at the University of California (Walker et al. 1994a) for controlling fanleaf degeneration (Walker et al. 1991). The O39-16 is resistant to most strains of root-knot nematodes (Lider et al. 1988a). It has a very strong resistance to California dagger nematode (*Xiphinema index*), the dagger nematode vector of grapevine fanleaf virus (GFLV). Although this rootstock resists *X. index* feeding, it allows the movement of GFLV into scions. However, the severe decline in the fruit set typical of GFLV infection does not occur. Due to O39-16's ability to alter phytohormone levels, O39-16 can induce tolerance to fanleaf degeneration. It can induce high vigor to scions where adequate water is available. It can produce wines with high pH and potassium levels and vegetative characters. This rootstock is a full sibling to O43-43 (FPS Grapes 2018).

(c) **VR O43-43 (043-43, OLMO VR O43-43)** (*V. vinifera* 'Almeria' × *V. rotundifolia* #1)

While it has been reported that O43-43 is tolerant to phylloxera (Lider et al. 1988b), some effects of phylloxera decline were observed in some areas of California (Walker 1991). O43-43 is offered for planting in vineyards infected with viruliferous nematodes (Lider et al. 1988b). It is a lime-susceptible rootstock (Bavaresco et al. 2005). This rootstock is a full sibling to VR O39-16 (FPS Grapes 2018).

(d) **VR O44-4 (044-4, OLMO VR O44-4)** (*V. vinifera* × *V. rotundifolia*)

The O44-4 shows a good level of resistance in field trials that appear to type A and B phylloxera (Granett et al. 1987). It induced high potassium to magnesium ratio in the petioles of scions (Dalbó et al. 2011).

(e) **Howell** (*V. vinifera* × *V. rotundifolia*)

This rootstock was developed by Michigan State University and has roots that peneterate to the deep soil, and is efficient in uptake of water and nutrients required for better growth of the scion cultivar (Howell 1987; Walker 1991).

11.20.8 *V. vinifera* × *V. rufotomentosa*

(a) **L171-6 (171-6)** (*V. vinifera* 'French Colombard' × *V. rufotomentosa*)

It is recommended for resistance to *Xiphinema index* nematode feeding and fanleaf viruses. However, ELISA testing was reported positive for grapevine fanleaf viruses (GFLV) (Walker et al. 1994a). L171-16 maintained its ability to induce vegetative growth while infected and produced the greatest pruning weights (Walker et al. 1994b).

11.20.9   V. vinifera × V. rupestris

(a)   **1202C (1202 Couderc)** (*V. vinifera* 'Mourvèdre'(Mataro) × *V. rupestris* 'Ganzin')

Roots show many tuberosities and are not fully phylloxera resistant. However, vines seem to grow well in spite of the damage (Galet 2004; Pool et al. 2005). The lack of phylloxera resistance indicates it should not be used as a rootstock where phylloxera is present. It has some salt and lime tolerance (Pongracz 1983) and is especially recommended for highly calcareous and deep sandy soils (Galet 2004; Pool et al. 2005). It is a very vigorous rootstock (Pongracz 1983).

(b)   **A × R 1 (A × R1, A × R#1, AXR1, AXR #1, Ganzin 1, ARG 1)** (*V. vinifera* 'Aramon' × *V. rupestris* 'Ganzin')

It has been grown widely but has failed to maintain enough phylloxera resistance over time (Pongracz 1983; Galet 2004). Therefore, it should not be planted (Davis 2016) and is no longer recommended as a rootstock because of the susceptibility to phylloxera (FPS Grapes 2018). All populations of phylloxera capable of feeding aggressively on AXR1 have been termed biotype B (Davis 2016). It has some relative tolerance to most viruses and good lime tolerance (Galet 2004; Davis 2016). AXR1's advantages were high vigor, adaptability, consistent high quality and yields, and easy of propagation (Davis 2016).

(c)   **A × R 9 (A × R#9, AXR#9, AXR9)** (*V. vinifera* Aramon × *V. rupestris* 'Ganzin')

It shows a low phylloxera resistance rating (Boubals 1966) and is not recommended as rootstock for phylloxera susceptibility (Wolpert and Walker 2002).

(d)   **G1 (G1, Grézot 1, Grezot 1)** (derived from a 1202 Couderc seedling (*V. vinifera* 'Mourvèdre' × *V. rupestris*))

G1 presents moderate resistance to radicicola phylloxera, and must be used under unfavorable conditions for this pest. It has moderate resistance to gallicola phylloxera and good resistance to downy mildew. It is susceptible to chlorosis and resists just up to 6% of active limestone. It is also susceptible to drought. This rootstock easily absorbs magnesium and is quite resistant to magnesium deficiency but has difficulty absorbing potassium in the soil. It is also susceptible to chlorides and not recommended for soil with the risk of salinity. G1 is a moderate to high vigor rootstock and generally shows a good affinity to grafts. This rootstock produces good results with 'Chasselas'. This rootstock is a low to moderate wood producer, but it shows good cuttings and grafting capacity (INRA 2016).

11.20.10   *V. vinifera* × *V. rupestris* × *V. cordifolia*

(a)   **62-66 C (62-66 Couderc)** (3102 Couderc (*V. vinifera* 'Colombeau' × *V. rupestris* 'Martin') × *V. cordifolia*)

It is resistant to *Meloidogyne javanica* (Clingeleffer and Smith 2011), and it is reported that it shows a very good rooting and grafting capacity (Samarakis 1924).

11.20.11   *V. vinifera* × *V. rupestris* × *V. riparia*

(a)   **196-17Cl (196-17 Castel)** (1203 Couderc (*V. vinifera* 'Monastrell' × *V. rupestris* 'Ganzin') × *V. riparia* 'Gloire de Montpellier')

196-17Cl shows a moderate tolerance to radicicolae phylloxera, but has good tolerance to the gallicola phylloxera. It is sensitive to *Meloidogyne arenaria* and *Meloidogyne incognita* nematodes. It shows moderate resistance to downy mildew. 196-17Cl is sensitive to chlorosis and resists just up to 6% of active limestone and a CPI of 5. It is well adapted to acid soils and shows a low tolerance to chlorides. 196-17Cl is adapted to drought and schist or granite, dry, lean, superficial, and stony soils along with sandy soils. It has good vigor and can be used for replants of lacking plants. It is a moderate wood producer but shows a good cutting and grafting capacity (INRA 2016).

11.20.12   *V. vinifera* × *V. solonis* × *V. riparia*

(a)   **Sorisil** (Silvaner (*V. vinifera*) × 1616 E Couderc ((*V. solonis* × *V. riparia*))

It is a phylloxera-resistant rootstock (Becker et al. 2003). It shows a medium to high chlorosis tolerance and high drought tolerance (Pavloušek 2013). It is adapted to various soils and different microclimates. It presents a good grafting affinity with different grape varieties (Kaserer and Schoffl 1993).

11.20.13   *V. vinifera* × *V. vinifera*

(a)   **Autumn Black** (*V. vinifera* Calmeria × *V. vinifera* Blackrose)

Autumn Black is a late-maturing black seeded table grape (*Vitis vinifera* L.) with excellent fruit quality following long-term storage (Ramming 1987), which is sometimes used as a rootstock (FPS Grapes 2018).

11.20.14   V. vinifera × V. vinifera × V. rotundifolia × V. berlandieri × V. rupestris

(a)	**NAB (Nemadex Alain Bouquet)** (crossbreeding of the hybrid F1 (VMH 8771) of *V. vinifera* ('Cabernet Sauvignon' × 'Alicante Henri Bouschet') × *V. rotundifolia* (NC 184-4) × 140 Ru (*V. berlandieri* 'Rességuier 2' × *V. rupestris* 'du Lot'))

NAB has a very high tolerance to radicicola phylloxera, and is not very susceptible to gallicola phylloxera. It presents a partial resistance to the nematodes *Xiphinema index* and which in the vineyards delays infection and slows down fan-leaf virus contamination. NAB resists up to 15% of active limestone and a CPI of 30. It has low to moderate resistance to ferric chlorosis. NAB shows low resistance to drought and also is sensitive to temporary excess humidity in the spring. Its vigor is quite low with bushy upright growth and short internodes. Consequently, it is a very low wood producer but has a moderate to good cuttings and grafting capacity. NAB produces moderate yields (INRA 2016).

References for Figures

Fig. 11.1 Blade shapes of mature leaves of commercial grape rootstocks

5A-Martinez Zaporta (http://www.vivc.de)

Ursula Brühl, Julius Kühn-Institut (JKI), Federal Research Centre for Cultivated Plants, Institute for Grapevine Breeding Geilweilerhof—76,833 Siebeldingen, GERMANY-Var#VIVC7448

5A-Teleki (http://www.vivc.de)

Julius Kühn-Institut (JKI), Federal Research Centre for Cultivated Plants, Institute for Grapevine Breeding Geilweilerhof—76,833 Siebeldingen, GERMANY-Var#VIVC12323

26 G (http://www.vivc.de)

Julius Kühn-Institut (JKI), Federal Research Centre for Cultivated Plants, Institute for Grapevine Breeding Geilweilerhof—76,833 Siebeldingen, GERMANY-Var#VIVC4634

34 EM (http://www.vivc.de)

Ursula Brühl, Julius Kühn-Institut (JKI), Federal Research Centre for Cultivated Plants, Institute for Grapevine Breeding Geilweilerhof—76,833 Siebeldingen, GERMANY-Var#VIVC3815

41 B Mgt (http://www.vivc.de)

Julius Kühn-Institut (JKI), Federal Research Centre for Cultivated Plants, Institute for Grapevine Breeding Geilweilerhof—76,833 Siebeldingen, GERMANY-Var#VIVC7736

44-53 Malègue (http://www.vivc.de)

Ursula Brühl, Julius Kühn-Institut (JKI), Federal Research Centre for Cultivated Plants, Institute for Grapevine Breeding Geilweilerhof—76,833 Siebeldingen, GERMANY-Var#VIVC7171

62-66 C (http://www.vivc.de)

Ursula Brühl, Julius Kühn-Institut (JKI), Federal Research Centre for Cultivated Plants, Institute for Grapevine Breeding Geilweilerhof—76,833 Siebeldingen, GERMANY-Var#VIVC2933

99 R (http://www.vivc.de)

Julius Kühn-Institut (JKI), Federal Research Centre for Cultivated Plants, Institute for Grapevine Breeding Geilweilerhof—76,833 Siebeldingen, GERMANY-Var#VIVC10064

101-14 Mgt (http://www.vivc.de)

Julius Kühn-Institut (JKI), Federal Research Centre for Cultivated Plants, Institute for Grapevine Breeding Geilweilerhof—76,833 Siebeldingen, GERMANY-Var#VIVC7745

106-8 Mgt (http://www.vivc.de)

Ursula Brühl, Julius Kühn-Institut (JKI) Bundesforschungsinstitut für Kulturpflanzen Institut für Rebenzüchtung Geilweilerhof—76,833 Siebeldingen—GERMANY-Var#VIVC7749

Fig. 11.2 Blade shapes of mature leaves of commercial grape rootstocks

110 R (http://www.vivc.de)

Julius Kühn-Institut (JKI), Federal Research Centre for Cultivated Plants, Institute for Grapevine Breeding Geilweilerhof—76,833 Siebeldingen, GERMANY-Var#VIVC10065

140 Ru (http://www.vivc.de)

Ursula Brühl, Julius Kühn-Institut (JKI), Federal Research Centre for Cultivated Plants, Institute for Grapevine Breeding Geilweilerhof—76,833 Siebeldingen, GERMANY-Var#VIVC10351

143A Mgt (http://www.vivc.de)

Doris Schneider, Ursula Brühl, Julius Kühn-Institut (JKI), Federal Research Centre for Cultivated Plants, Institute for Grapevine Breeding Geilweilerhof—76,833 Siebeldingen, GERMANY-Var#VIVC7768

157-11 C (http://www.vivc.de)

Ursula Brühl, Julius Kühn-Institut (JKI) Bundesforschungsinstitut für Kulturpflanzen Institut für Rebenzüchtung Geilweilerhof—76,833 Siebeldingen—GERMANY-Var#VIVC42288

160-19 C (http://www.vivc.de)

Julius Kühn-Institut (JKI), Federal Research Centre for Cultivated Plants, Institute for Grapevine Breeding Geilweilerhof—76,833 Siebeldingen, GERMANY-Var#VIVC3015

161-49 C (http://www.vivc.de)

Ursula Brühl, Julius Kühn-Institut (JKI) Bundesforschungsinstitut für Kulturpflanzen Institut für Rebenzüchtung Geilweilerhof—76,833 Siebeldingen—GERMANY-Var#VIVC 3016

188-04 Cl (http://www.vivc.de)

Ursula Brühl, Julius Kühn-Institut (JKI), Federal Research Centre for Cultivated Plants, Institute for Grapevine Breeding Geilweilerhof—76,833 Siebeldingen, GERMANY-Var#VIVC2167

188-15 Cl (http://www.vivc.de)

Ursula Brühl, Julius Kühn-Institut (JKI), Federal Research Centre for Cultivated Plants, Institute for Grapevine Breeding Geilweilerhof—76,833 Siebeldingen, GERMANY-Var#VIVC2169

196-17 Cl (http://www.vivc.de)

Ursula Brühl, Julius Kühn-Institut (JKI), Federal Research Centre for Cultivated Plants, Institute for Grapevine Breeding Geilweilerhof—76,833 Siebeldingen, GERMANY-Var#VIVC 2170

216-3 Cl (http://www.vivc.de)

Ursula Brühl, Julius Kühn-Institut (JKI) Bundesforschungsinstitut für Kulturpflanzen Institut für Rebenzüchtung Geilweilerhof—76,833 Siebeldingen—GERMANY-Var#VIVC 2175

Fig. 11.3 Blade shapes of mature leaves of commercial grape rootstocks

333 EM (http://www.vivc.de)

Ursula Brühl, Julius Kühn-Institut (JKI), Federal Research Centre for Cultivated Plants, Institute for Grapevine Breeding Geilweilerhof—76,833 Siebeldingen, GERMANY-Var#VIVC 12,520

420 A Mgt (http://www.vivc.de)

Ursula Brühl, Julius Kühn-Institut (JKI), Federal Research Centre for Cultivated Plants, Institute for Grapevine Breeding Geilweilerhof—76,833 Siebeldingen, GERMANY-Var#VIVC 7810

554-5 C (http://www.vivc.de)

Ursula Brühl, Julius Kühn-Institut (JKI), Federal Research Centre for Cultivated Plants, Institute for Grapevine Breeding Geilweilerhof—76,833 Siebeldingen, GERMANY-Var#VIVC3103

775 P (http://www.vivc.de)

Ursula Brühl, Julius Kühn-Institut (JKI), Federal Research Centre for Cultivated Plants, Institute for Grapevine Breeding Geilweilerhof—76,833 Siebeldingen, GERMANY-Var#VIVC9007

779 P (http://www.vivc.de)

Ursula Brühl, Julius Kühn-Institut (JKI), Federal Research Centre for Cultivated Plants, Institute for Grapevine Breeding Geilweilerhof—76,833 Siebeldingen, GERMANY-Var#VIVC9008

1045 P (http://www.vivc.de)

Ursula Brühl, Julius Kühn-Institut (JKI), Federal Research Centre for Cultivated Plants, Institute for Grapevine Breeding Geilweilerhof—76,833 Siebeldingen, GERMANY-Var#VIVC9018

1103 P (http://www.vivc.de)

Ursula Brühl, Julius Kühn-Institut (JKI), Federal Research Centre for Cultivated Plants, Institute for Grapevine Breeding Geilweilerhof—76,833 Siebeldingen, GERMANY-Var#VIVC 9023

1202 C (http://www.vivc.de)

Ursula Brühl, Julius Kühn-Institut (JKI), Federal Research Centre for Cultivated Plants, Institute for Grapevine Breeding Geilweilerhof—76,833 Siebeldingen, GERMANY-Var#VIVC3123

1447 P (http://www.vivc.de)

Ursula Brühl, Julius Kühn-Institut (JKI), Federal Research Centre for Cultivated Plants, Institute for Grapevine Breeding Geilweilerhof—76,833 Siebeldingen, GERMANY-Var#VIVC 9037

1613 C (http://www.vivc.de)

Julius Kühn-Institut (JKI), Federal Research Centre for Cultivated Plants, Institute for Grapevine Breeding Geilweilerhof—76,833 Siebeldingen, GERMANY-Var#VIVC3131

Fig. 11.4 Blade shapes of mature leaves of commercial grape rootstocks

1616 C (http://www.vivc.de)

Julius Kühn-Institut (JKI), Federal Research Centre for Cultivated Plants, Institute for Grapevine Breeding Geilweilerhof—76,833 Siebeldingen, GERMANY-Var#VIVC 3134

3306 C (http://www.vivc.de)

Julius Kühn-Institut (JKI), Federal Research Centre for Cultivated Plants, Institute for Grapevine Breeding Geilweilerhof—76,833 Siebeldingen, GERMANY-Var#VIVC3157

3309 C (http://www.vivc.de).

Ursula Brühl, Julius Kühn-Institut (JKI) Bundesforschungsinstitut für Kulturpflanzen Institut für Rebenzüchtung Geilweilerhof—76,833 Siebeldingen—GERMANY-Var#VIVC 3160

4010 Cl (http://www.vivc.de)

Ursula Brühl, Julius Kühn-Institut (JKI), Federal Research Centre for Cultivated Plants, Institute for Grapevine Breeding Geilweilerhof—76,833 Siebeldingen, GERMANY-Var#VIVC 2212

6736 Cl (http://www.vivc.de)

Ursula Brühl, Julius Kühn-Institut (JKI), Federal Research Centre for Cultivated Plants, Institute for Grapevine Breeding Geilweilerhof—76,833 Siebeldingen, GERMANY-Var#VIVC2237

AXR#1 (http://www.vivc.de)

Ursula Brühl, Julius Kühn-Institut (JKI), Federal Research Centre for Cultivated Plants, Institute for Grapevine Breeding Geilweilerhof—76,833 Siebeldingen, GERMANY-Var#VIVC4393

Beta (http://www.vivc.de)

Ursula Brühl, Julius Kühn-Institut (JKI) Bundesforschungsinstitut für Kulturpflanzen Institut für Rebenzüchtung Geilweilerhof—76,833 Siebeldingen—GERMANY-Var#VIVC1291

Binova (http://www.vivc.de)

Julius Kühn-Institut (JKI), Federal Research Centre for Cultivated Plants, Institute for Grapevine Breeding Geilweilerhof—76,833 Siebeldingen, GERMANY-Var#VIVC1369

Börner (http://www.vivc.de)

Julius Kühn-Institut (JKI), Federal Research Centre for Cultivated Plants, Institute for Grapevine Breeding Geilweilerhof—76,833 Siebeldingen, GERMANY-Var#VIVC1499

Champanel (http://www.vivc.de)

Doris Schneider, Julius Kühn-Institut (JKI), Federal Research Centre for Cultivated Plants, Institute for Grapevine Breeding Geilweilerhof—76,833 Siebeldingen, GERMANY-Var#VIVC2440

Fig. 11.5 Blade shapes of mature leaves of commercial grape rootstocks

Clinton (http://www.vivc.de)

Ursula Brühl, Julius Kühn-Institut (JKI), Federal Research Centre for Cultivated Plants, Institute for Grapevine Breeding Geilweilerhof—76,833 Siebeldingen, GERMANY-Var#VIVC2711

Cosmo 2 (http://www.vivc.de)

Ursula Brühl, Julius Kühn-Institut (JKI), Federal Research Centre for Cultivated Plants, Institute for Grapevine Breeding Geilweilerhof—76,833 Siebeldingen, GERMANY-Var#VIVC2866

Cosmo 10 (http://www.vivc.de)

Ursula Brühl, Julius Kühn-Institut (JKI), Federal Research Centre for Cultivated Plants, Institute for Grapevine Breeding Geilweilerhof—76,833 Siebeldingen, GERMANY-Var#VIVC2870

Dog Ridge (http://www.vivc.de)

Ursula Brühl, Julius Kühn-Institut (JKI), Federal Research Centre for Cultivated Plants, Institute for Grapevine Breeding Geilweilerhof—76,833 Siebeldingen, GERMANY-Var#VIVC3623

Du Lot (http://www.vivc.de)

Doris Schneider, Julius Kühn-Institut (JKI), Federal Research Centre for Cultivated Plants, Institute for Grapevine Breeding Geilweilerhof—76,833 Siebeldingen, GERMANY-Var#VIVC10389

Evex 13-5 (http://www.vivc.de)

Julius Kühn-Institut (JKI), Federal Research Centre for Cultivated Plants, Institute for Grapevine Breeding Geilweilerhof—76,833 Siebeldingen, GERMANY-Var#VIVC4011

Fercal (http://www.vivc.de)

Ursula Brühl, Julius Kühn-Institut (JKI) Bundesforschungsinstitut für Kulturpflanzen Institut für Rebenzüchtung Geilweilerhof—76,833 Siebeldingen— GERMANY-Var#VIVC4087

Freedom (http://www.vivc.de).

Ursula Brühl, Julius Kühn-Institut (JKI), Federal Research Centre for Cultivated Plants, Institute for Grapevine Breeding Geilweilerhof—76,833 Siebeldingen, GERMANY-Var#VIVC4235.

Gravesac (http://www.vivc.de).

Ursula Brühl, Julius Kühn-Institut (JKI), Federal Research Centre for Cultivated Plants, Institute for Grapevine Breeding Geilweilerhof—76,833 Siebeldingen, GERMANY-Var#VIVC931

Fig. 11.6 Blade shapes of mature leaves of commercial grape rootstocks

Grézot 1 (http://www.vivc.de)

Ursula Brühl, Julius Kühn-Institut (JKI), Federal Research Centre for Cultivated Plants, Institute for Grapevine Breeding Geilweilerhof—76,833 Siebeldingen, GERMANY-Var#VIVC5015

GRN-1 (Walker, 2009)

Walker, M.A., 2009, Nematod-resistant rootstocks released to nurseries, Practical Winery & Vineyard, Grape Growing, January/February 2009, 3p

GRN-2 (Walker, 2009)

Walker, M.A., 2009, Nematod-resistant rootstocks released to nurseries, Practical Winery & Vineyard, Grape Growing, January/February 2009, 3p

GRN-3 (Walker, 2009)

Walker, M.A., 2009, Nematod-resistant rootstocks released to nurseries, Practical Winery & Vineyard, Grape Growing, January/February 2009, 3p

GRN-4 (Walker, 2009)

Walker, M.A., 2009, Nematod-resistant rootstocks released to nurseries, Practical Winery & Vineyard, Grape Growing, January/February 2009, 3p

GRN-5 (Walker, 2009)

Walker, M.A., 2009, Nematod-resistant rootstocks released to nurseries, Practical Winery & Vineyard, Grape Growing, January/February 2009, 3p

Harmony (http://www.vivc.de)

Ursula Brühl, Julius Kühn-Institut (JKI), Federal Research Centre for Cultivated Plants, Institute for Grapevine Breeding Geilweilerhof—76,833 Siebeldingen, GERMANY-Var#VIVC5309

IAC313 (http://www.infobibos.com/Artigos/2010_2/PortaEnxertos/index.htm)

José Luiz Hernandes, J.L., F.P. Martins, and M.J.P. Júnior, 2010, Uso de porta enxertos—Tecnologia simples e fundamental na cultura da videira, Infobibos, Informações Tecnológicas

IAC 571–6 (http://www.infobibos.com/Artigos/2010_2/PortaEnxertos/index.htm

José Luiz Hernandes, J.L., F.P. Martins, and M.J.P. Júnior, 2010, Uso de porta enxertos—Tecnologia simples e fundamental na cultura da videira, Infobibos, Informações Tecnológicas

IAC 572 (http://www.infobibos.com/Artigos/2010_2/PortaEnxertos/index.htm)

José Luiz Hernandes, J.L., F.P. Martins, and M.J.P. Júnior, 2010, Uso de porta enxertos—Tecnologia simples e fundamental na cultura da videira, Infobibos, Informações Tecnológicas.

IAC 766 (http://www.infobibos.com/Artigos/2010_2/PortaEnxertos/index.htm).

José Luiz Hernandes, J.L., F.P. Martins, and M.J.P. Júnior, 2010, Uso de porta enxertos—Tecnologia simples e fundamental na cultura da videira, Infobibos, Informações Tecnológicas

Jacquez (http://www.vivc.de)

Julius Kühn-Institut (JKI), Federal Research Centre for Cultivated Plants, Institute for Grapevine Breeding Geilweilerhof—76,833 Siebeldingen, GERMANY-Var#VIVC5627

Kingfisher (https://fps.ucdavis.edu/WebSitePDFs/Newsletters&Publications/FPS GrapeNewsletter2011.pdf)

Cousins, 2011, Three Root-Knot Nematode Resistant Rootstocks Released By USDA Agricultural Research Service, FPS Grapes Program Newsletter October 2011, pp 12

Kober 5 BB (http://www.vivc.de)

Ursula Brühl, Julius Kühn-Institut (JKI), Federal Research Centre for Cultivated Plants, Institute for Grapevine Breeding Geilweilerhof—76,833 Siebeldingen, GERMANY-Var#VIVC 6313

Fig. 11.7 Blade shapes of mature leaves of commercial grape rootstocks

Kober 125 AA (http://www.vivc.de)

Julius Kühn-Institut (JKI), Federal Research Centre for Cultivated Plants, Institute for Grapevine Breeding Geilweilerhof—76,833 Siebeldingen, GERMANY-Var#VIVC12344

Lake Emerald (http://www.vivc.de)

Ursula Brühl, Julius Kühn-Institut (JKI), Federal Research Centre for Cultivated Plants, Institute for Grapevine Breeding Geilweilerhof—76,833 Siebeldingen, GERMANY-Var#VIVC6673

LN 33 (http://www.vivc.de)

Ursula Brühl, Julius Kühn-Institut (JKI), Federal Research Centre for Cultivated Plants, Institute for Grapevine Breeding Geilweilerhof—76,833 Siebeldingen, GERMANY-Var#VIVC6885

M1   (http://www.assoenologi.it/main/images/pics/relazione_scienza_71congresso.pdf)

Atilio Scienza, 2016, La Sfida Della ricerca Genetica per Contrastare Gli Effetti Del Cambiamento Climatico E Dei Parassiti Della Vite, 71 congresso Nazionale, Verona 2/5 Gugno 2016

M2   (http://www.assoenologi.it/main/images/pics/relazione_scienza_71congresso.pdf)

Atilio Scienza, 2016, La Sfida Della ricerca Genetica per Contrastare Gli Effetti Del Cambiamento Climatico E Dei Parassiti Della Vite, 71 congresso Nazionale, Verona 2/5 Gugno 2016

M3   (http://www.assoenologi.it/main/images/pics/relazione_scienza_71congresso.pdf)

Atilio Scienza, 2016, La Sfida Della ricerca Genetica per Contrastare Gli Effetti Del Cambiamento Climatico E Dei Parassiti Della Vite, 71 congresso Nazionale, Verona 2/5 Gugno 2016

M4   (http://www.assoenologi.it/main/images/pics/relazione_scienza_71congresso.pdf)

Atilio Scienza, 2016, La Sfida Della ricerca Genetica per Contrastare Gli Effetti Del Cambiamento Climatico E Dei Parassiti Della Vite, 71 congresso Nazionale, Verona 2/5 Gugno 2016

Matador (https://fps.ucdavis.edu/WebSitePDFs/Newsletters&Publications/FPSGrapeNewsletter2011.pdf)

Cousins, 2011, Three Root-Knot Nematode Resistant Rootstocks Released By USDA Agricultural Research Service, FPS Grapes Program Newsletter October 2011, pp 12

NAB (http://www.vivc.de)

Ursula Brühl, Julius Kühn-Institut (JKI) Bundesforschungsinstitut für Kulturpflanzen Institut für Rebenzüchtung Geilweilerhof—76,833 Siebeldingen—GERMANY-Var#VIVC23334

Fig. 11.8 Blade shapes of mature leaves of commercial grape rootstocks

Ramsey (http://www.vivc.de)

Julius Kühn-Institut (JKI), Federal Research Centre for Cultivated Plants, Institute for Grapevine Breeding Geilweilerhof—76,833 Siebeldingen, GERMANY-Var#VIVC9904

RGM (http://plantgrape.plantnet-project.org/en/porte-greffes)

Pl@ntGrape, Catalogue of Vines Cultivated in France, IFV—INRAE—l'Institut Agro | Montpellier SupAgro, 2009–2021"

© UMT Géno-Vigne® INRA—IFV—Montpellier SupAgro edited on 11/02/2021

Schwarzman (http://www.vivc.de)

Ursula Brühl, Julius Kühn-Institut (JKI), Federal Research Centre for Cultivated Plants, Institute for Grapevine Breeding Geilweilerhof—76,833 Siebeldingen, GERMANY-Var#VIVC10836

SO4 (http://www.vivc.de)

Julius Kühn-Institut (JKI), Federal Research Centre for Cultivated Plants, Institute for Grapevine

Breeding Geilweilerhof—76,833 Siebeldingen, GERMANY-Var#VIVC11473

Sori (http://www.vivc.de)

Julius Kühn-Institut (JKI), Federal Research Centre for Cultivated Plants, Institute for Grapevine Breeding Geilweilerhof—76,833 Siebeldingen, GERMANY-Var#VIVC11926

Sorisil (http://www.vivc.de)

Julius Kühn-Institut (JKI), Federal Research Centre for Cultivated Plants, Institute for Grapevine Breeding Geilweilerhof—76,833 Siebeldingen, GERMANY-Var#VIVC11927

Teleki 5 C (http://www.vivc.de)

Julius Kühn-Institut (JKI), Federal Research Centre for Cultivated Plants, Institute for Grapevine Breeding Geilweilerhof—76,833 Siebeldingen, GERMANY-Var#VIVC12324

Teleki 8 B (http://www.vivc.de)

Julius Kühn-Institut (JKI), Federal Research Centre for Cultivated Plants, Institute for Grapevine Breeding Geilweilerhof—76,833 Siebeldingen, GERMANY-Var#VIVC12326.

References

Alexandrov E (2017) Determination of the resistance to environmental factors of the interspecific hybrids of vines (*Vitis vinifera* L. × *Muscadinia rotundifolia* Michx.). Agricultura-Ştiinţă şi practică 1(2):101–102

Alleweldt G, Possingham JV (1988) Progress in grapevine breeding. Theor Appl Genet 75(5):669–673

Anders F (1958) Aminosauren als gallenerregende Stoffe der Reblaus (*Viteus* [Phylloxera] *vitifoloii* Shimer). Experientia 14:62–63

Anwar SA, McKenry M, Ramming D (2002) A search for more durable grape rootstock resistance to root-knot nematode. Am J Enol Vitic 53(1):19–23

Bavaresco L, Gardiman M, Brancadoro L, Espen L, Failla O, Scienza A, Vezzulli S, Zulini L, Velasco R, Stefanini M, Di Gaspero G, Testolin R (2015) Grapevine breeding programs in Italy. In: Reynolds A (ed) Grapevine breeding programs for the wine industry. Woodhead Publishing Series in Food Science, Technology and Nutrition: Number 268

Bavaresco L, Presutto P, Civardi S (2005) VR O43-43: a lime-susceptible rootstock. Am J Enol Vitic 56:192–195

Becker A, Schartl A, Herrmann JV (2003) Newly bred varieties of phylloxera tolerant rootstocks. Acta Hortic 603:745–748. https://doi.org/10.17660/ActaHortic.2003.603.105

Becker H, Schmid J, Ries R (2010) Grapevine varieties and site conditions in the viticultural regions of Hesse. Geologische Abhandlungen Von Hessen 114:70–79

Bettiga L, Golina D, McGourty G, Smith R, Verdegaal P, Weber E (2003) Wine grape varieties in California. UC ANR publication 3419. https://anrcatalog.ucanr.edu/Details.aspx?itemNo=3419

Bioletti FT (1908) Grape culture in California. Bulletin No. 197, Berkeley, Superintendent State printing 115–158

Börner C (1943) Die ersten reblausimmunen Rebenkreuzungen. Angew Bot 25:126–143

Boubals D (1966) A study of the distribution and causes of resistance to radicicolous phylloxera in the Vitaceae. Annales D'amelioration Des Plantes 16(2):145–183

Cirami R (1994) The J Stocks—two nematode-resistant [grape] rootstocks which feature low pH and low potassium with satisfactory yield. Australian Grapegrower Winemaker J 336(a):126–127

Clarke AD (2004) Suggested rootstocks for New Zealand vineyards, Ben & Frances Wickham Ormond Nurseries Ltd, pp. 1-8. https://ormondnurseries.co.nz/cms/uploads/pdf/rootstock_report.pdf

Clingeleffer P, Davis H (2012) Rootstock breeding and development for Australian dried grapes. CSIRO Plant Industry, Horticulture Australia Ltd., Canberra

Clingeleffer P, Smith B (2011) Rootstock breeding and development for Australian wine grapes, CSIRO plant industry, final report to grape and wine research & development corporation, CSP 05/03, 102 p

Cosmo I, Comuzzi A, Polsinelli M (1958) Portnnesti della vite ediziom agricole. Bologna. Italy, pp 27–45

Cousins P (2012) Three root-knot nematode resistant rootstocks released by USDA-ARS. In: Vasquez SJ (ed) Vine Lines, Fresno County, University of California, pp 1–9. http://cefresno.ucdavis.edu

Credi R, Silvestroni O, Santucci A, Canova A (1991) Variation in grapevine rootstock susceptibility to the rugose wood disease "legno riccio." Phytopathol Mediterr 30(1):41–46

Dalbó MA, Feldberg N (2016) Agronomic behavior of grape rootstocks resistant to young vine decline in Santa Catarina State, Brazil. In: BIO web of conferences 7, 01017 (2016), 39th world congress of vine and wine, pp 1–3. https://doi.org/10.1051/bioconf/20160701017

Dalbó MA, Schuck E, Basso C (2011) Influence of rootstock on nutrient content in grape petioles. Rev Bras Frutic Jaboticabal—SP 33(3):941–947

Daly A, Hennessy C (2007) Grapevine leaf rust project, final report, grape and wine research and development corporation, Australian Government, 36 p

Davis UC (2016) Viticultural information. University of California, UC Integrated Viticulture http://iv.ucdavis.edu/Viticultural_Information, Accessed on 26 Nov 2016

Dias FAN, da Mota RV, de Souza CR, de Azevedo Pimentel RM, de Souza LC, de Souza AL, de Albuquerque Regina M (2017) Rootstock on vine performance and wine quality of 'Syrah' under double pruning management. Sci Agric 74(2):134–141

Dry N (2005) Final report to grape and wine research and development corporation. Phylloxera and Grape Industry Board of South Australia, PGI 05/01, 22 p

Dry N (2008) Grapevine rootstocks. Selection and management for South Australia vineyards. Phylloxera and Grape Industry Board of South Australia. Lythrum Press, 85 p

Elkins R, Walker A (1991) Grape rootstock characteristics. In: Lake County horticultural notes, 8 p. http://celake.ucanr.edu/newsletters/September-October_199128631.pdf. Accessed on 15 Oct 2019

Embrapa (2014a) IAC 313. https://www.embrapa.br/en/uva-e-vinho/cultivares-e-porta-enxertos/porta-enxertos/-/asset_publisher/rE0HjHq6jP8J/content/porta-enxerto-iac-313-tropical/1355300. Accessed on 17 Sept 2019

Embrapa (2014b) IAC 572. https://www.embrapa.br/en/uva-e-vinho/cultivares-e-porta-enxertos/porta-enxertos/-/asset_publisher/rE0HjHq6jP8J/content/porta-enxerto-iac-572-jales/1355300. Accessed on 17 Sept 2019

Embrapa (2014c) IAC 766. https://www.embrapa.br/uva-e-vinho/cultivares-e-porta-enxertos/porta-enxertos/-/asset_publisher/rE0HjHq6jP8J/content/porta-enxerto-iac-766-campinas/1355300. Accessed on 17 Sept 2019

Ferreira JHS (1985) Effect of rootstock on the incidence of dying arm of Chenin blanc vines. S Afr J Enol Vitic 6(1):23–24

Ferris H, Zheng L, Walker MA (2012) Resistance of grape rootstocks to plant-parasitic nematodes. J Nematol 44(4):377–386

Fisher H (2021) Using novel grapevine rootstocks for stabilizing yield in winter injury prone wine districts. OGWRI project # 002600, Interim report 3, 13 p

FPS Grapes (2018) Foundation plant services grapes http://fps.ucdavis.edu/fgrsearch.cfm. Accessed on Jan 2019

Galet P (2004) Cépages et vignobles de France. Tome 3: Les vignobles de France, vol 1

Gee CT, Gadoury DM, Cadle-Davidson L (2008) Ontogenic resistance to *Uncinula necator* varies by genotype and tissue type in a diverse collection of *Vitis* spp. Plant Disease 92(7):1067–1073

Geisenheim (2016) Grape rootstocks. Hochschule Geisenheim University, Center for Applied Biology, Institute for Grapevine Breeding. https://www.hs-geisenheim.de/en/research/departments/grapevine-breeding/department-of-grapevine-breeding. Accessed on 26 Nov 2016

Granett J, Goheen AC, Lider LA, White JJ (1987) Evaluation of grape rootstocks for resistance to type A and type B grape phylloxera, Am J Enol Vitic 38(4):298-300.

Hayne P (1896) Resistant vines, their selection, adaptation, and grafting, Appendix to viticultural report, University of California, College of Agriculture, Agricultural experiment station

Hennessy CR, Daly AM, Hearnden MN (2007) Assessment of grapevine cultivars for resistance to *Phakopsora euvitis*. Australas Plant Pathol 36:313–317

Hidalgo L, Candela MR (1979) 5A Martinez-Zaporta (5A-MZ), New grapevine rootstock, Instituto Nacional de Investigaciones Agrarias, Madrid (Spain).

Howell GS (1987) Vitis rootstocks. In: Rom RC, Carlson RF (eds) Rootstocks for fruit crops, Wiley Inc., New York, pp 451–475

Howell GS (2005) Rootstock influence on scion performance. Grapevine rootstocks; current use, research and application. In: Proceedings of the 2005 rootstock symposium. Pub by MVEC, pp 47–55

IAC, Instituto Agronomico Campinas (2019) Cultivares IAC, São Paulo, Brazil. http://www.iac.sp.gov.br/cultivares/inicio/resultados.php. Accessed on 25 Sept 2019

INRA (2016) Plant Grape. UMT Geno-Vigne, INRA-IFV-Montepellier SupAgro, http://plantgrape.plantnet-project.org/en/porte-greffes. Accessed on 26 Nov 2016

Jahnke G, Molnár GK, Májer J, Szőke B, Tarczal E, Varga P, Kocsis L (2011) Analysis of grape rootstocks by SSR markers. Journal International Des Sciences De La Vigne Et Du Vin 45(3):1–12

Justia Patents (1994) Grape plant named 'Borner'. https://patents.justia.com/patent/PP9575. Accessed on 17 Sept 2019

Justia Patents (2003a) Grape plant named 'RS-3'. https://patents.justia.com/patent/PP16291. Accessed on 17 Sept 2019

Justia Patents (2003b) Grape plant named 'RS-9'. https://patents.justia.com/patent/PP16115. Accessed on 17 Sept 2019

Kamas J (2014) Growing grapes in Texas: from the commercial vineyard to the backyard vine, Texas A&M University Press, 264 p

Kaserer H, Schoffl G (1993) Ergebnisse eines langjahrigen versuches uber die affinitat rebsorte zweigelt zu den unterlagen 5C, S04, 5BB, 143A, Sorisil und G9. Mitteilungen Klostemeuburg 43(4):109–117

Kubota N, Li XG, Yasui K (1993) Effects of rootstocks on sugar, organic acid, amino acid, and anthocyanin contents in berries of potted 'Fujiminori' grapes. J Japan Soc Hort Sci 62:363–370

Laguna Lumbreras E (2004a) American and hybrid grapevines (Vitis spp.): A new concept of invasive plants to Europe. In: Proceedings of the 4th European conference on the conservation of the wild plants, A workshop on the implementation of the global strategy for plant conservation in Europe, Valencia, Spain. https://www.researchgate.net/figure/Fig-n-1-Representation-of-leaves-of-several-pure-species-of-Vitis-L-most-of-them_fig1_267960737

Laguna Lumbreras E (2004b) Datos foliares de las especies e híbridos alóctondos de vides (Género Vitis) en el territorio valenciano. https://www.researchgate.net/publication/229150974

Laucou V, Boursiquot JM, Lacombe T, Bordenave L, Decroocq S, Ollat N (2008) Parentage of grapevine rootstock 'Fercal' finally elucidated. Vitis 47(3):163–167

Liang Z, Yang Y, Cheng L, Zhong G (2013) Characterization of polyphenolic metabolites in grape hybrids. Vitis 52(2):51–59

Lider LA, Olmo HP, Goheen AC (1988a) The Regents of the University of California, assignee. Hybrid grapevine rootstock named 'VR O39-16'. U.S. plant patent 6, 166

Lider LA, Olmo HP, Goheen AC (1988b) The Regents of the University of California, assignee. Hybrid grapevine rootstock named 'VR O43-43'. U.S. plant patent 6, 319

Marais PG (1983) Phytophthora cinnamomic root rot of grapevines in South Africa, University of Stellenbosch, PhD dissertation. 122 p

McCarthy MG, Cirami RM, Furkaliev DG (2008) Rootstock response of Shiraz (Vitis vinifera) grapevines to dry and drip-irrigated conditions. Aust J Grape Wine Res 3(2):95–98

McEachern GR, Stein L, Kamas J (1997) Growing pierce's disease resistant grapes in Central, South and East Texas, https://aggie-horticulture.tamu.edu/extension/fruit/Piercegrapes/pdr.html. Accessed on 12 June 2019

Micheloni C (2017) EIP-AGRI focus group, diseases and pests in viticulture. Starting paper, EIP-AGRI service point. Version 19 January 2017, 18 p

Mortensen JA (1980) Source and inheritance of resistance to Anthracnose (Elsinoe ampelina) in Vitis. In: Proceeding of the third international symposium on grape breeding, Davis, pp 263–274

Mortensen JA, Harris JW, Hopkins DL (1994) Florilush': a bunch grape rootstock for Florida. HortScience 29(11):1375–1376

Mortensen JA, Stover LH (1982) Tampa—a new bunch grape rootstock. Agricultural Experiment Stations, University of Florida, Circular S-295, pp 1–4

Mortensen JA, Stover LH, Balerdi CF (1977) Source of resistance to Pierce's disease in Vitis. J Am Soc Horticult Sci 102:695–697

Niklowitz W (1955) Histologische studien an reblausgallen und reblausabwehrnekrosen. Phytopathol Z 24:299–340

Nogueira DJP (1984) Porta-Enxertos De Videiras. Informe Agropecuário 10:22–24

Ortiz JM, Cabello F, Muñoz G (2012) Status report of Vitis germplasm in Spain, pp 1–8. http://archive-ecpgr.cgiar.org/fileadmin/www.ecpgr.cgiar.org/NW_and_WG_UPLOADS/VITIS_2_2012/SPAIN.pdf

Owens CL (2012) Grape rootstock. In: Finn CE, Clark JR (eds). Register of new fruit and nut cultivars. List 46, USDA-ARS, Grape Genetics Research Unit, Geneva. Hortscience 47(5):545–546

Pàstena B (1976) Trattato di Viticoltura Italian. Edagricola, Bologna, pp 379–403

Pathirana R, McKenzie MJ (2005) Early detection of grapevine leafroll virus in Vitis vinifera using in vitro micrografting. Plant Cell Tiss Organ Cult 81:11–18. https://doi.org/10.1007/s11240-004-2498-y

Pavloušek P (2012) Screening of rootstock hybrids with *Vitis cinerea* Arnold for phylloxera resistance. Cent Eur J Biol 7(4):708–719. https://doi.org/10.2478/s11535-012-0062-z

Pavloušek P (2013) Tolerance to lime—induced chlorosis and drought in grapevine rootstocks. In: Vahdati K, Leslie C (eds) Abiotic stress: plant responses and applications in agriculture, Chap 9, pp 277–306, IntechOpen

Perry RL, Sabbatini P (2015) Grape rootstocks for Michigan. Michigan State University, Department of Horticulture, E3298:1–8 (www.grapes.msu.edu)

Plantgrape (2019) Catalogue of vines grown in France, INRA - IFV - Montpellier SupAgro, Retrieved 16 Sep 2019. http://plantgrape.plantnet-project.org/en/cepage/Jaoumet

Pommer CV (2000) Cultivares de uva produzidos ou introduzidos pelo IAC. Simpósio Brasileiro Sobre Uvas De Mesa 1:51–67

Pongrácz DP (1983) Rootstocks for grapevines. David Philip, Cape Town, South Africa

Pool B, Lerch S, Howard G, Johnson T, Weimann D (2005) Rootstocks for planting or replanting New York vineyards, Finger Lakes Vineyard Notes. Cornell University, pp 1–14

Pool RM (2004) Grapevine physiology and management of New York vineyards, A final report represented to USDA, https://portal.nifa.usda.gov/web/crisprojectpages/0006449-grapevine-phy siology-and-management-of-new-york-vineyards.html

Pool RM, Reisch BI, Welser MJ (1990) Use of differential thermal analysis to quantify bud cold hardiness of grape selections and clones, In: Proceeding fifth international symposium on grape breeding. Bundesforschungsanstalt für Rebenzüchtung, Geilweilerhof, Germany. (Vitis special issue), pp 318–329

Rahemi A, Dale A, Fisher II, Taghavi T, Bonnycastle A, Kelly J (2016) A report on *Vitis riparia* in Ontario, Canada. Acta Hortic 1136:33–38

Ramming DW (1987) 'Autumn black' grape. Fruit Var J 41:90–91. https://www.pubhort.org/aps/41/v41_n3_a4.htm

Read PE, Gu S (2003) A century of American viticulture, HortScience 38(5):943-951.

Reynolds A (2015) Grapevine breeding programs for the wine industry, Woodhead Publishing Series in Food Science, Technology and Nutrition: Number 268

Robinson J, Harding J (2016) The oxford companion to wine, the, 4th edn. Oxford University Press

Ruehl E, Schmid J, Eibach R, Töpfer R (2015) Grapevine breeding programmes in Germany, In: Reynolds A (ed) Grapevine breeding programs for the wine industry. Woodhead Publishing series in food science, technology and nutrition: number 268, pp 77–101

Samarakis BA (1924) Bulletin de l'office international du Vin. p23.

Shao-Hua L (2001) Grape production in China, In: Papademetriou MK, Dent FJ (eds) Grape production in the Asia-Pacific region. FAO publication, Expert consultation on "Viticulture (grape production) in Asia and the Pacific", held in Bangkok, Thailand, 2–4 May 2000

Šimek M (2017) Rootstock varieties, book of Czech varieties, http://winegrapevarieties.blog.cz/rub rika/odrudy-podnozove-kniha-odrud-cr. Accessed on 23 Sept 2019

Smith BP, Morales NB, Thomas MR, Smith HM, Clingeleffer PR (2017) Grapevine rootstocks resistant to the root-knot nematode *Meloidogyne javanica*. Aust J Grape Wine Res 23:125–131

Southey JM, Jooste JH (1991) The effect of grapevine rootstock on the performance of *Vitis vinifera* L. (cv. Colombard) on a relatively saline soil. S Afr J Enol Vitic 12(1):32–41

Stobbs LW, van Schagen JG (1987) Effect of shoot thinning on the production of graftable wood from four grape rootstocks. Can J Plant Sci 67:289–291

Stover LH (1960) Blue Lake—a new bunch grape for Florida home gardens, University of Florida, Circular S-120, pp 1–9

Tanne E, Shlamovitz N, Spiegel-Royl P (1993) Rapidly diagnosing grapevine corky bark by in vitro micrografting. HortScience 28(6):667–668

Teubes A (2014) History of rootstocks in South Africa, Practical in the vineyard, Wineland Media, http://www.wineland.co.za/wp-content/uploads/2017/01/andrew4.jpg. Accessed on 14 Dec 2019

Tóth H, Kocsis L (2003) Level of phylloxera resistance of two new rootstock hybrids: Georgikon 28 and Georgikon 121. Acta Hortic 603:739–743

USDA, United States Department of Agriculture (1943) Grapes for different regions, Farmers Bulletin No (1936) USDA. Government Printing Office, Washington D.C. US, p 38

Vanden Heuvel JE, Proctor JTA, Sullivan JA, Fisher KH (2004) Influence of training/trellising system and rootstock selection on productivity and fruit composition of Chardonnay and Cabernet franc Grapevines in Ontario. Canada, Am J Enol Vitic 55(3):253–264

van Schalkwyk D (2017) Fruit production: Get the right rootstock for planting grapes, https://www.africanfarming.com/fruit-production-get-right-rootstock-planting-grapes. Accessed on 10 Sept 2019

VCR, Vivai Cooperativi Rauscedo (2007) Vitis general catalogue. http://www.vitisrauscedo.it/pdf/catalogue-en.pdf. Accessed on 16 Sept 2019

VCR, Vivai Cooperativi Rauscedo (2019) Overview of the characteristics of the vine rootstocks. http://www.vivairauscedo.com/en/portinnesti. Accessed on 16 Sept 2019

Viala P, Ravaz L (1903) American vines: their adaptation, culture, grafting and propagation. R Dubois and EH Twight, translators. Freygang-Leary Co., San Francisco. 299 p

Villa F, Pio R, Chalfun NNJ, Gontijo TCA, Coelho JHC, Dutra LF (2003) Rooting of herbaceous cuttings of the grape rootstock Riparia de Traviú treated with auxins. Ciênc Agrotec Lavras 27(6):1426–1431

VIVC, Vitis International Variety Catalogue (2019) Passbord data, Utilization: Rootstock. http://www.vivc.de

Walker MA (1991) Characteristics of *Vitis* species commonly used in rootstock breeding. ASEV Publication, Davis, CA, Alternative Rootstock Update, pp 1–7

Walker MA (2009) Nematode-resistant rootstocks released to nurseries, Practical Winery & Vineyard, Grape Growing, January/February 2009, p3.

Walker MA, Lider LA, Goheen AC, Olmo HP (1991) VR 039–16 grape rootstock. HortScience 26(9):1224–1225

Walker MA, Wolpert JA, Weber E (1994) Field screening of grape rootstock selections for resistance to Fanleaf degeneration. Am Phytopathol Soc Plant Dis 78(2):134–136

Walker MA, Wolpert JA, Webber E (1994) Viticultural characteristics of VR hybrid rootstocks in a vineyard site infected with grapevine fanleaf virus. Vitis 33:19–23

Walker R, Blackmore D, Clingeleffer P, Walker MA, Gilliham M, Henderson S, Francis L, Holt H, Pearson W (2014) Delivering chloride and sodium excluding rootstocks for quality wine production. Grape and wine research and development corporation (Wine Australia), CSIRO, final report project number CSP1002

Whiting JR (2003) Selection of grapevine rootstocks and clones for Greater Victoria. Department of Primary Industries: Victoria, 40 p

Whiting JR, Buchanan GA (1992) Evaluation of rootstocks for phylloxera infested vineyards in Australia. In: Rootstock seminar. A worldwide perspective. Am Soc Enol Vitic, pp 15–26

Wine Growers (2019) Rootstocks for grafted vines. http://www.winegrowers.info/rootstocks/home.htm

Wolpert J, Walker MA (2002) Field evaluation of winegrape rootstocks, Final report to American vineyard foundation, California rootstock commission, Viticulture Consortium, 112 p

Correction to: Abiotic Stresses and Rootstock Selection

Correction to:
Chapter 6 in: A. Rahemi et al., *Grape Rootstocks and Related Species*, https://doi.org/10.1007/978-3-030-99407-5_6

The original version of the book was inadvertently published with an extra figure in Chapter 6, which has now been removed. The chapter and the book have been updated with the change.

The updated version of this chapter can be found at
https://doi.org/10.1007/978-3-030-99407-5_6

A. Rahemi et al., *Grape Rootstocks and Related Species*,
https://doi.org/10.1007/978-3-030-99407-5_12

Appendix 1
The Plant List, Vitis, Royal Botanic Gardens

The Plant List (2013) Version 1.1. Published on the Internet <http://www.theplantl ist.org/> (accessed August 2019).

Vitis species list			
Scientific name	Subgenous	Groups	Status
Vitis acerifolia Raf. (Including *V. solonis* Hort. Berol ex Planchon and *V. longii* Prince & Prince, Laguna 2004a)	Euvitis	American species	Accepted
Vitis aestivalis Michx	Euvitis	American species	Accepted
Vitis amazonica (Linden) G. Nicholson	Euvitis	American species	Accepted
Vitis amurensis Rupr	Euvitis	Asian species	Accepted
Vitis arizonica Engelm	Euvitis	American species	Accepted
Vitis baileyana Munson	Euvitis	American species	Accepted
Vitis balansana Planch	Euvitis	Asian species	Accepted
Vitis bashanica P.C.He	Euvitis	Asian species	Accepted
Vitis bellula (Rehder) W.T. Wang	Euvitis	Asian species	Accepted
Vitis berlandieri Planch	Euvitis	American species	Accepted
Vitis betulifolia Diels & Gilg	Euvitis	Asian species	Accepted
Vitis biformis Rose	Euvitis	American species	Accepted
Vitis bloodwothiana Comeaux	Euvitis	American species	Accepted
Vitis bourgaeana Planch	Euvitis	American species	Accepted
Vitis bryoniifolia Bunge	Euvitis	Asian species	Accepted
Vitis californica Benth	Euvitis	American species	Accepted
Vitis candicans Engelm. ex A. Gray	Euvitis	American species	Unresolved
Vitis × champinii Planch. (*V. candicans* × *V. rupestris* (Goldammer 2018) or *V. mustangensis* × *V. rupestris* USDA)	Euvitis	American species	Unresolved

(continued)

(continued)

Vitis species list			
Scientific name	Subgenous	Groups	Status
Vitis chontalensis Seem	Euvitis	American species	Accepted
Vitis chunganensis Hu	Euvitis	Asian species	Accepted
Vitis chungii F.P. Metcalf	Euvitis	Asian species	Accepted
Vitis cinerea (Engelm.) Engelm. ex Millardet	Euvitis	American species	Accepted
Vitis cissoides (Blume) Backer	Euvitis	Asian species	Accepted
Vitis coignetiae Pulliat ex Planch	Euvitis	Asian species	Accepted
Vitis cordifolia Lam	Euvitis	American species	Accepted
Vitis davidii (Rom.Caill.) Foëx	Euvitis	Asian species	Accepted
Vitis enneaphylla (Vell.) Eichler	Euvitis	Asian species	Accepted
Vitis erythrophylla W.T. Wang	Euvitis	Asian species	Accepted

Scientific name	Sub genous	Groups	Status
Vitis fengqinensis C.L. Li	Euvitis	Asian species	Accepted
Vitis figariana (Webb) Baker	Euvitis	Asian species	Accepted
Vitis flexuosa Thunb	Euvitis	Asian species	Accepted
Vitis girdiana Munson	Euvitis	American species	Accepted
Vitis hancockii Hance	Euvitis	Asian species	Accepted
Vitis heyneana Roem. & Schult	Euvitis	Asian species	Accepted
Vitis hui W.C. Cheng	Euvitis	Asian species	Accepted
Vitis jacquemontii R. Parker	Euvitis	Asian species	Accepted
Vitis jinggangensis W.T. Wang	Euvitis	Asian species	Accepted
Vitis jinzhainensis X.S. Shen	Euvitis	Asian species	Accepted
Vitis labrusca L	Euvitis	American species	Accepted
Vitis labruscana L.H. Bailey	Euvitis	American species	Accepted
Vitis lanceolatifoliosa C.L. Li	Euvitis	Asian species	Accepted
Vitis linsecomii Buckley	Euvitis	American species	Accepted
Vitis longii Prince and Prince	Euvitis	American species	Unresolved
Vitis longquanensis P.L. Chiu	Euvitis	Asian species	Accepted
Vitis luochengensis W.T. Wang	Euvitis	Asian species	Accepted
Vitis menghaiensis C.L. Li	Euvitis	Asian species	Accepted
Vitis mengziensis C.L. Li	Euvitis	Asian species	Accepted
Vitis monticola Buckley	Euvitis	American species	Accepted
Vitis munsoniana J.H. Simpson ex Planch. Syn. *Muscadinia munsoniana* (J.H. Simpson ex Planch.) Small	Muscadinia	American species	Accepted
Vitis mustangensis Buckley	Euvitis	American species	Accepted

(continued)

(continued)

Scientific name	Sub genous	Groups	Status
Vitis nesbittiana Comeaux	Euvitis	American species	Accepted
Vitis × *novae-angliae* Fernald (pro sp.)	Euvitis	American species	Accepted
Vitis papillosa (Blume) Backer	Euvitis	Asian species	Accepted
Vitis piasezkii Maxim	Euvitis	Asian species	Accepted
Vitis pilosonerva F.P. Metcalf	Euvitis	Asian species	Accepted
Vitis popenoei J.L. Fennell., Syn. *Muscadinia popenoei* (J.L. Fennell) Weakley and Gandhi	Muscadinia	American species	Accepted

Scientific name	Sub genous	Groups	Status
Vitis pseudoreticulata W.T. Wang	Euvitis	Asian species	Accepted
Vitis pubescens (Schltdl.) Miq	Euvitis	American species	Accepted
Vitis retordii Rom. Caill. ex Planch	Euvitis	Asian species	Accepted
Vitis rhomboidea (E. Mey. ex Harv.) Szyszy	Euvitis	Asian species	Accepted
Vitis romanetii Rom. Caill	Euvitis	Asian species	Accepted
Vitis rotundifolia Michx., Syn., *Muscadinia rotundifolia* (Michx.) Small	Muscadinia	American species	Accepted
Vitis rufotomentosa Small	Euvitis	American species	Accepted
Vitis rupestris Scheele	Euvitis	American species	Accepted
Vitis ruyuanensis C.L. Li	Euvitis	Asian species	Accepted
Vitis shenxiensis C.L. Li	Euvitis	Asian species	Accepted
Vitis shuttleworthii House	Euvitis	American species	Accepted
Vitis silvestrii Pamp	Euvitis	Asian species	Accepted
Vitis simpsonii Munson	Euvitis	American species	Accepted
Vitis sinocinerea W.T. Wang	Euvitis	Asian species	Accepted
Vitis sinuata (Pursh) G. Don	Euvitis	Asian species	Accepted
Vitis thunbergii Siebold & Zucc	Euvitis	Asian species	Accepted
Vitis tiliifolia Humb. and Bonpl. ex Schult., Syn. *Vitis caribaea* De Candolle	Euvitis	American species	Accepted
Vitis treleasei Munson ex L.H. Bailey	Euvitis	American species	Accepted
Vitis tsoi Merr	Euvitis	Asian species	Accepted
Vitis tsukubana (Makino) Maekawa	Euvitis	Asian species	Accepted
Vitis unifoliata (Harv.) Kuntze	Euvitis	Asian species	Accepted
Vitis vinifera L	Euvitis	Eurasian species	Accepted
Vitis vulpina L., Syn. *Vitis riparia* Michx	Euvitis	American species	Accepted
Vitis wenchowensis C. Ling	Euvitis	Asian species	Accepted
Vitis wilsoniae H.J. Veitch	Euvitis	Asian species	Accepted
Vitis wuhanensis C.L. Li	Euvitis	Asian species	Accepted

(continued)

(continued)

Scientific name	Sub genous	Groups	Status
Vitis yuenlingensis W.T. Wang	Euvitis	Asian species	Accepted
Vitis zhejiang-adstricta P.L. Chiu	Euvitis	Asian species	Accepted

Appendix 2

Identification Key of North American Grape Species

An identification key for grape species in North America has illustrated on http://www.efloras.org: Flora of North America (*Vitis* identification key) by Moore and Wen (2016).

Moore MO, Wen J (2016) Vitaceae, in Flora of North America, North of Mexico, Vol. 12, eds Flora of North America Editorial Committee (New York, NY; Oxford: Oxford University Press).

1	Tendrils unbranched; bark adherent; lenticels prominent; pith continuous through nodes [1a. subg. *Muscadinia*]	1 *Vitis rotundifolia*
+	Tendrils branched or absent; bark exfoliating; lenticels inconspicuous or absent; pith interrupted by nodal diaphragms [1b. subg. *Vitis*]	(2)
2 (1)	Flowers bisexual; berries oblong to ellipsoid, skin adhering to pulp	2 *Vitis vinifera*
+	Flowers functionally unisexual; berries globose, skin separating from pulp	(3)
3 (2)	Leaf blade abaxial surface glaucous (sometimes obscured by hairs)	3 *Vitis aestivalis*
+	Leaf blade abaxial surface not glaucous (concealed by hairs in *V. labrusca*)	(4)
4 (3)	Tendrils or inflorescences present at 3+ consecutive nodes or almost all nodes	(5)
+	Tendrils or inflorescences present at only 2 consecutive nodes	(6)
5 (4)	Leaf blade abaxial surface densely and persistently arachnoid, concealed (except sometimes veins) by hairs; nodal diaphragms 0.5–2.5 mm thick; tendrils at almost all nodes	4 *Vitis labrusca*

(continued)

(continued)

+	Leaf blade abaxial surface ± densely arachnoid when young, sparsely arachnoid when mature, visible through hairs; nodal diaphragms 0.3–1.1 mm thick; tendrils usually not at all nodes	5 *Vitis* × *novae-angliae*
6 *(4)*	Leaf blade abaxial surface densely tomentose, concealed (except sometimes veins) by hairs; berries 12+ mm diam	*(7)*
+	Leaf blade abaxial surface usually glabrous, moderately arachnoid, or hirtellous, sometimes tomentose (California and s Oregon), visible through hairs; berries 4–12 mm diam. (except 12+ mm in *V.* × *doaniana* and *V.* × *champinii*)	*(8)*
7 *(6)*	Stipules 1.5–4 mm; nodal diaphragms 1.5–3 mm thick; Alabama, Arkansas, Louisiana, Oklahoma, Texas	6 *Vitis mustangensis*
+	Stipules to 1 mm; nodal diaphragms 2.5–6 mm thick; Florida	7 *Vitis shuttleworthii*
8 *(6)*	Leaf blades reniform, abaxial surface usually glabrous, sometimes sparsely hirtellous on veins and in vein axils; tendrils absent or only at distalmost nodes	8 *Vitis rupestris*
+	Leaf blades usually cordate to cordate ovate, sometimes orbiculate or nearly reniform, abaxial surface glabrous or hairy; tendrils along length of branchlets	*(9)*
9 *(8)*	Nodal diaphragms to 0.5(–1) mm thick; branchlet growing tips enveloped by unfolding leaves	*(10)*
+	Nodal diaphragms 1–4 mm thick; branchlet growing tips not enveloped by unfolding leaves	*(11)*
10 *(9)*	Plants low to moderately high climbing, much branched; tendrils soon deciduous if not attached to support; branchlets arachnoid or glabrate, growing tips sparsely to densely hairy; inflorescences 3–7(–9) cm	9 *Vitis acerifolia*
+	Plants usually moderate to high climbing, sometimes sprawling, sparsely branched; tendrils persistent; branchlets glabrous or sparsely hirtellous, growing tips glabrous or sparsely hairy; inflorescences (4–)9–12 cm	10 *Vitis riparia*
11 *(9)*	Berries 12+ mm diam	*(12)*
+	Berries 4–12 mm diam	*(13)*
12 *(11)*	Leaf blade abaxial surface moderately to densely arachnoid, hirtellous on veins; berries glaucous	11 *Vitis* × *doaniana*
+	Leaf blade abaxial surface sparsely arachnoid to glabrate, not hirtellous; berries usually not, sometimes very slightly, glaucous	12 *Vitis* × *champinii*
13 *(11)*	Leaf blade abaxial surface sparsely to densely tomentose; California, s Oregon	*(14)*
+	Leaf blade abaxial surface glabrous or sparsely to densely arachnoid or hirtellous; much of United States, but not California or Oregon	*(15)*

(continued)

(continued)

14 *(13)*	Berries moderately to heavily glaucous, 8–10 mm diam.; branchlet tomentum thinning in age; nodal diaphragms 3–4 mm thick	13 *Vitis californica*
+	Berries slightly or not glaucous, 4–6 mm diam.; branchlet tomentum usually persistent; nodal diaphragms 1.5–3 mm thick	14 *Vitis girdiana*
15 *(13)*	Plants sprawling to low climbing, shrubby, much branched; tendrils soon deciduous if not attached to means of support; Arizona, Nevada, New Mexico, trans-Pecos Texas, Utah	15 *Vitis arizonica*
+	Plants usually moderate to high climbing, sometimes ± shrubby and sprawling when without support, sparsely branched; tendrils persistent; e United States, including Texas, not in trans-Pecos region	*(16)*
16 *(15)*	Branchlets ± angled, densely hirtellous and/or sparsely to densely arachnoid, to glabrate; berries 4–8 mm diam.; nodes sometimes red-banded	16 *Vitis cinerea*
+	Branchlets terete or subterete, glabrous or sparsely arachnoid; berries 8–12 mm diam.; nodes not red-banded	*(17)*
17 *(16)*	Nodal diaphragms 2.5–4 mm thick; leaf blades usually deeply lobed, apices long acuminate; branchlets uniformly red, purplish red, or chestnut	17 *Vitis palmata*
+	Nodal diaphragms 1–2.5 mm thick; leaf blades unlobed or shallowly lobed, sometimes deeply lobed on ground shoots, apices acute to short acuminate; branchlets gray to green or brown, if purplish only on one side	*(18)*
18 *(17)*	Berries usually with lenticels; inflorescences 3–7 cm; branchlet growing tips sparsely to densely hairy; leaf blades 5–8(–10) cm; branchlets sparsely arachnoid or glabrous	18 *Vitis monticola*
+	Berries without lenticels; inflorescences 9–19 cm; branchlet growing tips glabrous to sparsely hairy; leaf blades (5–)9–18 cm; branchlets glabrous	19 *Vitis vulpine*

Appendix 3
Identification Key of Eastern North American Grape Species

One identification key was illustrated by Moore (1991), and considered 12 species and 9 varieties, distributed throughout the United States and Canada, largely east of the Rocky Mountains. Three hybrid taxa are also found in eastern North America (North of Mexico).

Moore MO (1991) Classification and systematics of eastern North American *Vitis* L. (Vitaceae), North of Mexico. Sida 14:339–367.

1. Tendrils simple; bark adherent with prominent lenticels; pith continuous through nodes → Subgenus Muscadinia.

1. Tendrils bifid to trifid; bark shedding, the lenticels inconspicuous; pith interrupted by diaphragms at nodes → 2 (Subgenus Vitis).

2. Leaves glaucous beneath; nodes often glaucous → Series Aestivales.

2. Leaves not glaucous beneath; nodes not glaucous → 3.

3. Branchlets of the season angled, pubescent with arachnoid or hirtellous trichomes, or both, varying to glabrate; mature 3 or 4 seeded berries less than 8 mm in diameter; nodes frequently banded with red pigmentation → Series Cinerescentes.

3. Branchlets of the season more or less terete, glabrous or pubescent; mature 3 or 4 seeded berries usually greater than 8 mm in diameter; nodes usually not banded with red pigmentation → 4.

4. Leaves heavily arachnoid pubescent beneath, concealing the leaf undersurface but not always the veins; mature fruits greater than 12 mm in diameter → Series Labruscae.

4. Leaves glabrous to slightly arachnoid pubescent beneath, not concealing the inter-vein area of the under-leaf surface; mature berries less than 12 mm in diameter → 5.

A. Rahemi et al., *Grape Rootstocks and Related Species*, https://doi.org/10.1007/978-3-030-99407-5

5. Growing tips more or less enveloped by enlarging, unfolding leaves; stipules large, usually greater than 3 mm long; nodal diaphragms usually less than 1 mm in diameter → Series Ripariae.

5. Growing tips not enveloped by enlarging, unfolding leaves; stipules small, usually less than 3 mm long; nodal diaphragms usually greater than I mm in diameter → Series Cordifoliae.

ARTIFICIAL KEY TO SPECIES AND HYBRIDS.

1. Tendrils simple; bark adherent with prominent lenticels; pith continuous through nodes → *V. rotundifolia.*

1. Tendrils bifid to trifid; bark shredding, the lenticels inconspicuous; pith interrupted by nodal diaphragms → 2.

2. Mature leaves glaucous beneath; nodes often glaucous → *V. aestivalis.*

2. Mature leaves nor glaucous beneath; nodes not glaucous → 3.

3. Tendrils or inflorescences present at three to many consecutive nodes4.

4. Leaves densely pubescent beneath, concealing the leaf under-surface but not always the veins; nodal diaphragms usually greater than 1 mm in diameter → *V. labrusca.*

4.Leaves slightly to moderately pubescent beneath, the leaf under-surface visible on mature leaves; nodal diaphragms usually less than 1 mm in diameter → *V. × novae-angliae.*

3. Tendrils or inflorescences present at only two consecutive nodes → 5.

5.Leaves densely pubescent beneath, concealing the leaf under-surfaces but not always the veins; fruits greater than 12 mm in diameter → 6.

6.Stipules greater than 1 mm long; leaves frequently concavely folded *V. mustangensis.*

6. Stipules less than 1 mm long; leaves not concavely folded → *V. shuttleworthii.*

5. Leaves glabrous to moderately pubescent beneath, the intervein area of leaf under-surfaces visible on mature leaves; fruits less or greater than 12 mm in diameter 7.

7. Leaves reniform, glabrous beneath at maturity; tendrils absent or present only opposite the uppermost nodes → *V. rupestris.*

7. Leaves cordate to cordate ovate, glabrous to pubescent beneath at maturity, tendrils present opposite most nodes → 8.

8. Nodal diaphragms less than 1 mm wide, usually less than 0.5 mm wide; growing tips enveloped by enlarging, un-folding leaves → 9.

9. Growing tips slightly to densely pubescent; branchlets of the season slightly to densely arachnoid pubescent; inflorescences usually less than 8 cm long *V. acerifolia*.

9. Growing tips glabrous to slightly pubescent; branchlets of the season usually lacking arachnoid pubescence; inflorescences usually greater than 8 cm long → *V. riparia*.

8. Nodal diaphragms greater than 1 mm wide; growing tips not enveloped by enlarging, unfolding leaves → 10.

10. Branchlets of the season angled, arachnoid and/or hirtellous pubescent, varying to glabrate; mature 3 or 4 seeded berries less than 8 mm in diameter; nodes frequently banded with red pigmentation → *V. cinerea*.

10. Branchlets of the season more or less terete, glabrous or arachnoid pubescent; mature 3 or 4 seeded berries usually greater than 8 mm in diameter; nodes usually not banded with red pigmentation → 11.

11. Mature 3 or 4 seeded berries greater than 12 mm in diameter;→ leaves arachnoid pubescent.

beneath → 12.

12. Leaves moderately to heavily arachnoid pubescent beneath, also with hirtellous trichomes along the veins; fruits glaucous *V. × doaniana*.

12. Leaves only slightly arachnoid pubescent.

beneath and lacking hirtellous trichomes; fruits not glaucous → *V. × champinii*.

11. Mature 3 or 4 seeded berries less than 12 mm in diameter; leaves usually lacking arachnoid pubescence beneath → 13.

13. Nodal diaphragms greater than 2.5 mm wide; leaf apices usually long acuminate; branchlets of the season with a purplish red cast → *V. palmata*.

13.Nodal diaphragms less than 2.5 mm wide; leaf apices usually acute to short acuminate; branchlets of the season gray to green or brown or with purplish pigmentation only on one side of the branchlet → 14.

14. Berriesusuallywithlenticels; in- fructescences with less than 12 berries; growing tips slightly to densely pubescent; leaf blades usually less than 8 cm long; branchlets of the season usually slightly arachnoid pubescent → *V. monticola*.

14. Berries without lenticels; infructescences with more than 12 berries; growing tips glabrous to slightly pubescent; leaf blades usually greater than 9 cm long; branchlets of the season Glabrous *V. Vulipina*

Two varieties of *V. rotundifolia* can be distinguished morphologically based on the following key:

1. Mature fruits greater than 12 mm in diameter; infructescences usually with less than 12 berries; leaf blades usually more than 5 cm in length → *V. rotundifolia* var. rotundifolia.

1. Mature fruits less than 12 mm in diameter; infructescences with more than 12 berries; leaf blades often less than 5 cm long → *V. rotundifolia* var. munsoniana.

The three varieties of *V. aestivalis* can be distinguished morphologically based on the following key:

1. Branchlets of the season heavily arachnoid pubescent; mature 3 or 4 seeded berries usually greater than 14 mm in diameter; stipules usually less than 1.5 mm long *V. aestivalis* var. lincecimii.

1. Branchlets of the season slightly to moderately arachnoid pubescent, or glabrous; mature 3 or 4 seeded berries usually less than 14 mm in diameter; stipules usually greater than 1.5 mm long 2.

2. Mature 3 or 4 seeded berries less than 9 mm in diameter; mature leaves glabrous to glabrate beneath; nodes usually glaucous; nodal diaphragms usually less than 2 mm in diameter → *V. aestivalis* var. bicolor.

2. Mature 3 or 4 seeded berries greater than 9 mm in diameter; mature leaves slightly to heavily arachnoid pubescent beneath; nodes usually not glaucous; nodal diaphragms usually greater than 2 mm in diameter → *V. aestivalis* var. aestivalis.

Four varieties of *V. cinerea* can be distinguished morphologically based on the following key:

1. Berries moderately to heavily glaucous; leaf blades glabrous to glabrate, usually less than 10 cm long; central Texas → *V. cinerea* var. helleri.

1. Berries only slightly to not glaucous; leaf blades pubescent, varying to glabrate, usually more than 10 cm long; e Texas east and northward → 2.

2. Branchlets of the season sparsely to densely hirtellous pubescent, often with arachnoid pubescence as well; leaf undersurfaces usually more or less uniformly hirtellous pubescent on veins → *V. cinerea* var. cinerea.

2. Branchlets of the season without evident hirtellous trichomes (if present, then concealed by arachnoid pubescence); leaf undersurfaces usually without hirtellous trichomes, or, when present, only very sparsely so → 3.

3. Branchlets slightly to densely arachnoid pubescent; nodes usually not banded with red pigmentation; leaves slightly to densely arachnoid pubescent beneath; Coastal Plain → *V. cinerea* var. ftoridana.

3. Branchlets glabrate to only slightly arachnoid pubescent; nodes usually banded with red pigmentation; leaves glabrous to very slightly arachnoid pubescent beneath; Piedmont and Mountains.→ *V. cinerea* var. baileyana.

Appendix 4
Identification Key of Grape Species in Southeastern US

An identification key for grape species in Southeastern United States has illustrated by Small (1903).

Small JK (1903) Flora of the Southeastern United States, New York, Published by the Author, pp 751–757.

Hypogynous disk present, either annular, cup-shaped or glanular: leaf-blades simple or ternately compounds.

Bark shreddy: pith interrupted by diaphragms at the nodes: tendrils forking. 1. Vitis.

Bark not shreddy: pith continuous through the nodes: tendrils simple. 2. Muscadina

1. **Vitis L.**: trailing or climbing vines, with a shreddy bark and forking tendrils. Pith interrupted at the nodes by diaphragms. leaves alternate: blades simple, palmately lobed, angled or coarsely toothed, petioled. Stipules usually small, caduceus. Flowers dioecious, polygamy-dioecious or rarely perfect, in elongated racemes, or panicles. Disk hypogynous, often glandular. Calyx minute. Corolla cadcaducous: petals cohering at the top like a cap. Stamens excerted, alternating with the lobes or glands of the disk. Ovary 2-celled or rarely 3–4-celled: styles short or conic. Ovules 2 in each cavity. berries juicy, mostly edible. Seeds few, pear-shaped with a constricted beak-like base. Grape.

Leaf-blades green and glabrous beneath at maturity or merely cobwebby about the nerves or in their axils.

Leaf-blades merely toothed, or sometimes angled or shallowly lobed.

    Low shrub.   1. *V. rupestris*

    Elongated trailing or climbing vines

        Diaphragms thin.

            Berries less than 14 mm. in diameter.   2. *V. monticola*

            Berries over 15 mm. in diameter.   3. *V. champinii*

        Diaphragms thick.

            Shoot terete, glabrous or early becoming so.

                Leaf-blades long-pointed, the teeth rather triangular, acute.   4. *V. cordifolia*

                Leaf-blades mostly rounded at the apex, the teeth rounded

                    and mucronate.   5. *V. helleri*

            Shoots angled, pubescent throughout the year.

                Leaf-blades not lustrous on either side: seeds 3-4 mm. long. 6. *V. baileyana*

                Leaf-blades lustrous above and often below: seeds

                    4-5 mm. long   7. *V. berladieri*

Leaf-blades prominently lobed.

    Lobes and sinuses of the leaf-blades acute: berries with a bloom.

        Young foliage glabrous or essentially so: petioles green.   8. *V. vulpina*

        Young foliage manifestly pubescen: petioles white-flloeculent.   9. *V. longii*

    Lobes of leaf-blades acuminate: sinuses obtuse: berries destitute of any bloom.   10. *V. palmata*

Leaf-blades densely woolly, or glaucous and sparingly pubescent beneath.

Lower surface of leaf-blades glaucous, glabrate.   11. *V. bicolor*
Lower surface of leaf-blades more or less densely woolly.

    Leaf-blades cobwebby or floeculent beneath at maturity.

        Young foliage gray-cobwebby or white-tomentose.

            Leaf-blades densely white-tomentose beneath.   12. *V. doaniana*

            Leaf-blades gray-cobwebby beneath.   13. *V. cinerea*

        Young foliage ferrugineous-pubescent, at least on the nerves of the leaf-blades.

            Twigs, and often the expanding leaves, clothed with a close felt-like pubescence.

                Berries mostly less than 10 mm. in diameter, not glaucous: Floridian

                  species.   14. *V. aestivalis*

                Berries mostly over 10 mm. in diameter, glaucous: Texan

                  species.   15. *V. linsecomii*

            Twigs, and expanding leaves, more or less densely tomentose.

                Twigs and petioles copiously tomentose: panicle-branches felty-

                  Floeculent during anthesis.

                  Leaf –blades not lobed, except on the shoots, the margins

                    shallowly sinuate-toothed.   16. *V. caribaea*

Leaf-blades usually manifestly lobed, the margins

coarsely toothed.   17. *V.rufotomentosa*

Twigs, and often petioles, glabrous or nearly so: panicle-branches

not felty-floeculent   18. *V. aestivalis*

Leaf-blades felt-like or densely tomentose beneath at maturity.

Tendrils, or flower-clusters, intermittent, usually every third node without tendril or inflorescence.

Leaf-blades densely tomentose with white hairs.

Berries over 12 mm. in diameter.   19. *V. candicans*

Berries less than 12 mm. in Diameter   20. *V. coriacea*

Leaf-blades densely tomentose with rusty or tawny hairs .   21. *V. simpsonii*

Tendrils continuous, usually every node with a tendril, or inflorescence. 22. *V. labrusca*

2. **Muscadina** Small: trailing or climbing vines, with a close bark and simple tendrils. Stems sometimes producing aerial roots. Pith Continuous, not Interrupted at the Nodes by Diaphragms. Leaves alternate: blades simple, angled or coarsely toothed, never densely pubescent. Stipules caduceus. Flowers polygamo-dioecious or dioecious, in alternate recemes or panicles. Disk hypoynous. Calyx minute. Corolla caduceus: petals cohering at the top as a cap, never expanding. Stamens exserted, alternating with the lobes or glands of the disk. Ovary mostly 2-celled. Styles stout. Ovules 2 in each cavity. Berry globose, finally edible. Seeds shaped like a coffee-grain. Bullace grape.

Berries 1–1.5 cm. in diameter; the skin and pulp tender, the latter acid: seeds 3–5 mm. long.→ 1. *M. munsoniana.*

Berries 1.5–2.5 cm. in diameter; the skin and pulp tough, the latter musky: seeds 6–9 mm, long.→ 2. *M. rotundifolia.*

Appendix 5
Identification Key of Grape Species in China

An identification key for grape species in china has reported by Ren and Wen (2007).

Ren H, Wen J (2007) Vitis. In: Wu Z-Y, Hong D-Y, Raven PH (eds) Flora of China 12. Science Press, Beijing and Missouri Botanical Garden Press, St. Louis, 210–222.

1a. Leaves palmately compound with 3–5 leaflets.

2a. Branchlets and inflorescence axes pilose; leaflets abaxially sparsely pubescent, or with woolly hairs when young, becoming sparsely so when mature→10. *V. piasezkii.*

2b. Branchlets and inflorescence axes with arachnoid tomentum; leaflets abaxially with persistent rubiginous arachnoid tomentum→37. *V. lanceolatifoliosa.*

1b. Leaves simple.

3a. Branchlets with prickles, becoming tuberculate on old branches→1. *V. davidii.*

3b. Branchlets and old branches without prickles or tubercles.

4a. Branchlets and petioles with setose and glandular hairs, glands with or without a stalk.

5a. Branchlets and petioles with dense glandular as well as simple hairs; leaves inconspicuously.

3–5-lobed or unlobed, margin with sharp teeth→2. *V. romanetii.*

5b. Branchlets and petioles with sparse glandular hairs; leaves conspicuously 3–5-lobed, margin with obtuse teeth→3. *V. shenxiensis.*

4b. Branchlets and petioles pilose or with arachnoid tomentum.

6a. Leaves abaxially with dense white or rubiginous arachnoid or woolly tomentum.

7a. Leaves 3–5-lobed or heterophyllous and mixed with undivided ones.

8a. Leaves 3–5-lobed, lobes broad.

9a. Branches with white woolly tomentum, erect hairs absent; tendrils bifurcate→28. *V. heyneana.*

9b. Branches pilose; tendrils unbranched or mixed with bifurcate ones→35. *V. sinocinerea.*

8b. Leaves 3–5-cleft or deeply parted, mixed with undivided ones.

10a. Leaves of two types, 3–5-cleft and undivided→31. *V. longquanensis.*

10b. Leaves all divided deeply or at least to middle, sometimes deep lobes pinnatifid, some lobes broad and undivided→36. *V. bryoniifolia.*

7b. Leaves undivided or inconspicuously 3–5-lobed.

11a. Leaves basally deeply cordate, notch rounded or acutely angular, sinuses close together or overlapping.

12a. Leaves oval or ovate-elliptic, abaxial veins with woolly tomentum→29. *V. retordii.*

12b. Leaves suborbicular or quinquangularly rounded, abaxial veins densely pubescent with appressed soft hairs→30. *V. menghaiensis.*

11b. Leaves basally slightly cordate or sub-truncate, sometimes mixed with conspicuously cordate ones.

13a. Branchlets and inflorescence axis ± with arachnoid tomentum→28. *V. heyneana.*

13b. Branchlets and inflorescence axis with variable hairiness (including arachnoid tomentum).

14a. Leaves adaxially with sparse arachnoid tomentum when young, then glabrescent; branchlets and inflorescence axis with sparse arachnoid tomentum→32. *V. bellula.*

14b. Leaves adaxially or only veins with dense, short, appressed hairs; inflorescence axis densely pubescent.

15a. Branchlets sparsely pilose; leaves slightly lobed, base truncate, only adaxial veins sparsely pubescent→33. *V. bashanica.*

15b. Branchlets densely pubescent; leaves undivided, base sub-cordate, adaxially densely pubescent→34. *V. hui.*

6b. Leaves abaxially glabrous, or pilose, or with sparse arachnoid tomentum.

16a. Leaves abaxially glabrous or only axils of veins with hairs, or young leaves with tomentum then falling off.

17a. Leaves sagittate or deltoid-sagittate, 3–5-lobed or cleft, or mixed with undivided ones; inflorescence narrow, small, cylindrical, with basal branches poorly developed.

18a. Leaf margin 12–16-toothed on each side, conspicuously ciliate, abaxial veinlets raised→17. *V. wenchowensis.*

18b. Leaf margin 7–10-toothed on each side, only upper part of margin with very short and sparse hairs, not ciliate, abaxial veinlets conspicuous, but not raised→20. *V. ruyuanensis.*

17b. Leaves variously shaped but neither sagittate nor deltoid-sagittate, undivided.

19a. Leaves ovate, oval, elliptic, or ovate-lanceolate, base slightly sub-cordate or sub-truncate.

20a. Leaves ovate or oval, not glaucous, veinlets inconspicuous→25. *V. flexuosa.*

20b. Leaves elliptic or ovate-lanceolate, usually glaucous, veinlets conspicuous and raised→8. *V. chungii.*

19b. Leaves oval, broadly ovate, or triangular-ovate, base cordate or deeply so.

21a. Basal sinuses close together or partly overlaid.

22a. Leaves sub-leathery, margin serrate, apex sharply acute, abaxially usually glaucous, rarely glaucescent→6. *V. chunganensis.*

22b. Leaves papery, margin coarsely dentate, apex cuspidate, abaxially not glaucous→21. *V. mengziensis.*

21b. Basal sinuses forming an obtuse angle without any overlapping.

23a. Peduncles with sparse arachnoid tomentum when young, then gradually falling off →4. *V. balansana.*

23b. Peduncles pilose.

24a. Leaves oval, apex acute or acuminate, veinlets slightly raised→ 5. *V. yunnanensis.*

24b. Leaves ovate-oblong or triangular-ovate, apex acuminate, veinlets conspicuously raised → 7. *V. luochengensis.*

16b. Leaves abaxially pilose or at least abaxial veins pilose or with arachnoid tomentum.

25a. Leaves conspicuously 3–5-divided or mixed with undivided leaves.

26a. Leaves basally deeply cordate, basal sinuses attaching or partly overlapping, margin coarsely toothed (cultivated)→ 27. *V. vinifera.*

26b. Leaves basally cordate, basal notch obtuse or rounded, margin shallowly toothed.

27a. Tendrils unbranched, rarely bifurcate.

28a. Branchlets and petioles sparsely brown villous; leaves undivided or mixed with 2- or 3-cleft leaves, margin with long ciliate hairs→23. *V. hancockii.*

28b. Branchlets, petioles, and peduncles sparsely pilose; leaves usually 3–5-lobed or cleft, margin not ciliate or only slightly so with short ciliate hairs.

29a. Leaves abaxially greenish or lavender, veinlets slightly raised→18. *V. jinggangensis.*

29b. Leaves abaxially reddish purple, veinlets conspicuously raised on both sides→19. *V. erythrophylla.*

27b. Tendrils 2(or 3)-branched.

30a. Leaves unlobed or 3–5-lobed, lobes broad.

31a. Leaves oval or obelliptic, basal sinuses forming an obtuse angle→9. *V. betulifolia.*

31b. Leaves broadly oval, basal sinuses rounded, rarely forming an obtuse angle→26. *V. amurensis.*

30b. Leaves 3–5-cleft to middle or deeply parted, lobes narrow, rarely lobes pinnatifid, or sometimes mixed with shallowly lobed leaves.

32a. Plants small, stems slender; leaf margin 5–9-toothed on each side→15. *V. silvestrii.*

32b. Plants strong, stems robust; leaf margin 16–25-toothed on each side→14. *V. zhejiang-adstricta.*

25b. Leaves undivided, rarely inconspicuously 3–5-lobed.

33a. Leaves abaxially with at least veins with arachnoid tomentum, rarely glabrous, never with erect hairs.

34a. Inflorescence axis pilose and with arachnoid tomentum→22. *V. fengqinensis.*

34b. Inflorescence axis with arachnoid tomentum, but not erect hairs.

35a. Leaves 7–16 × 5–12 cm, margin 16–20-toothed on each side→12. *V. wilsoniae.*

35b. Leaves 3–7 × 2.5–6 cm, margin 5–13-toothed on each side→16. *V. wuhanensis.*

33b. Leaves abaxially with at least veins with erect hairs, or sometimes mixed with arachnoid tomentum.

36a. Leaves rhombic-ovate or rhombic-elliptic, base broadly cuneate or suborbicular→23. *V. hancockii.*

36b. Leaves ovate or oval, base cordate, sub-cordate, or sub-truncate, rarely rounded.

37a. Leaves basally conspicuously cordate.

38a. Branchlets glabrous; abaxial veins of leaves pilose with spreading brown hairs→11. *V. pilosonerva.*

38b. Branchlets with sparse woolly tomentum; abaxial veins of leaves with white hairs and with sparse arachnoid tomentum→ 13. *V. pseudoreticulata.*

37b. Leaves basally sub-cordate or sub-truncate, rarely rounded.

39a. Tendrils bifurcate; leaves oval or ovate-elliptic→9. *V. betulifolia.*

39b. Tendrils unbranched; leaves ovate or oval-lanceolate→24. *V. tsoi.*

Printed in the United States
by Baker & Taylor Publisher Services